TECHNICAL REPORT

Army Tactical Wheeled Vehicles

Current Fleet Profiles and Potential Strategy Implications

Carolyn Wong • Louis R. Moore • Elvira N. Loredo • Aimee Bower
Brian Pascuzzi • Keenan D. Yoho

Prepared for the United States Army
Approved for public release; distribution unlimited

ARROYO CENTER

The research described in this report was sponsored by the United States Army under Contract No. W74V8H-06-C-0001.

Library of Congress Control Number: 2011923086

ISBN: 978-0-8330-5093-9

The RAND Corporation is a nonprofit research organization providing objective analysis and effective solutions that address the challenges facing the public and private sectors around the world. RAND's publications do not necessarily reflect the opinions of its research clients and sponsors.

RAND® is a registered trademark.

Published 2011 by the RAND Corporation
1776 Main Street, P.O. Box 2138, Santa Monica, CA 90407-2138
1200 South Hayes Street, Arlington, VA 22202-5050
4570 Fifth Avenue, Suite 600, Pittsburgh, PA 15213-2665
RAND URL: http://www.rand.org/
To order RAND documents or to obtain additional information, contact
Distribution Services: Telephone: (310) 451-7002;
Fax: (310) 451-6915; Email: order@rand.org

Preface

This report documents the results of a multipart research project on evolving a fleet strategy for Army tactical wheeled vehicles. The first phase of the study focused on the light tactical vehicle fleet. The second phase focused on the medium and heavy tactical vehicle fleets, specifically building a status profile that shows the Army where it currently stands in terms of the types, quantities, and years of useful life remaining in its medium and heavy tactical wheeled vehicle fleets.

In addition, we discuss the state of the data with regard to tactical vehicles and present recommendations for future research. This research should be of interest to members of the Army and the Department of Defense responsible for formulating, reviewing, or implementing policy that governs the planning and acquisition of tactical wheeled vehicles for the Army.

The light tactical wheeled vehicle research was sponsored by BG Charles Anderson, who at the time was Director, Force Development, Deputy Chief of Staff, G-8, U.S. Army. The medium and heavy tactical wheeled research was sponsored by MG David Halverson, Director, Force Development, Office of the Deputy Chief of Staff, G-8, U.S. Army, and Mr. Christopher Lowman, Deputy Chief of Staff, G-4. This research was conducted within RAND Arroyo Center's Force Development and Technology Program. RAND Arroyo Center, part of the RAND Corporation, is a federally funded research and development center sponsored by the United States Army.

The Project Unique Identification Code (PUIC) for the project that produced this document is ASPMO09163.

For more information on RAND Arroyo Center's Force Development and Technology Program, contact the Director, Bruce Held (telephone 310-393-0411, extension 7405; email Bruce_Held@rand.org; mail RAND Corporation, 1776 Main Street, Santa Monica, California 90407-2138).

For more information on RAND Arroyo Center, contact the Director of Operations (telephone 310-393-0411, extension 6419; FAX 310-451-6952; email Marcy_Agmon@rand.org), or visit Arroyo's website at http://www.rand.org/ard/.

Contents

Figures

Tables

Summary

The Army's medium and heavy tactical wheeled vehicle (TWV) fleets (both active and reserve components) are critical to sustaining its global operations: these are the vehicles that move supplies and equipment to and around the battlespace. The Army has maintained a significant program and made major investments in its medium and heavy TWV fleets because they are such critical assets. More than $16 billion (fiscal year 2009 dollars) have been invested over the last five years to procure medium and heavy TWVs. Nevertheless, the program has not been able to keep up with the demands of its aging fleets. Today there are medium and heavy vehicles that are over 30 years old. Perhaps more important, the pace and requirements of current operations, particularly in Iraq and Afghanistan, and predeployment training are stressing these fleets even more. Both the actual and imposed aging of these vehicles suggest that the Army needs to update its TWV strategy, a key element toward managing its investments prudently. In order to update the TWV strategy effectively, the Army must be able to make informed decisions about its investments in replacing, upgrading, and using its TWV fleets. Informed decisions are dependent on a clear understanding of how many of what types of vehicles the Army currently has and some indication of their age and condition, as well as what the Army's requirements are likely to be at points in the future.

This study produced status profiles of the Army's medium and heavy tactical wheeled vehicle fleets. The status profiles show how many medium and heavy TWVs of each type the Army has and the years of useful life remaining for each group.

The study team integrated diverse data elements supplied by a TWV Integrated Product Team (IPT) and other sources to construct a data base that could be used to generate status profiles of the medium and heavy TWV fleets.

For the purposes of this study, the expected useful life (EUL) of a vehicle is the time until it reaches a point of such extensive and widespread wear that it is more economical to recapitalize or replace the vehicle than to continue to maintain and repair it. The study team used the EUL concept to transform the status data base to an easily interpretable graph whereby vehicles with different EULs can be placed on a common timeline. In all, the study team determined the statuses of 40 models of heavy TWVs and 117 models of medium TWVs. The aggregate status profile of the Army's oldest vehicles is shown in Figure S.1. The overview status profile of the Army's medium and heavy TWV fleets is shown in Figure S.2.

This overview of the status of the medium and heavy TWV fleets (Figure S.2) shows that the Army is in the middle of a window of opportunity. That is, the Army has just entered a period where relatively few groups of medium and heavy TWVs are exceeding their EUL. An effective Army strategy would seek to exploit this window of opportunity to take care of those vehicles that have already exceeded EUL before the next wave hits in about five years. Although five years seems like a lengthy amount of time, that might only be an illusion. Pro-

Figure S.1
The Army's Oldest TWVs

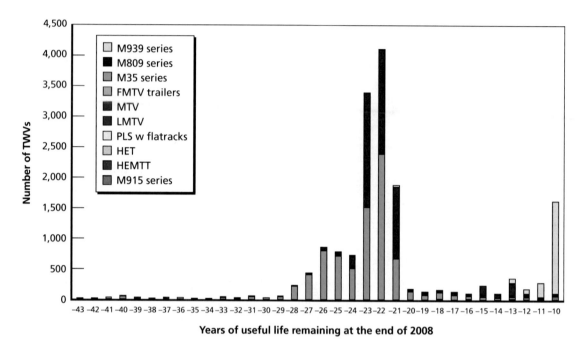

NOTE: Assumes 15-year EUL for medium TWVs and 20-year EUL for heavy TWVs.
RAND *TR890-S.1*

gram Objective Memorandum (POM) 2011 is near, and POM 2012 will soon command the Army's attention, so from a planning point of view, the Army's real window of opportunity is closer to two or three years. Hence, to take maximum advantage of the current window of opportunity, the Army needs to initiate immediate efforts to devise a TWV strategy that will serve it well into the future. The graph in Figure S.2 indicates a likely starting point to base the Army's updated TWV strategy.

This study has provided the Army with starting points for the medium and heavy TWV fleets, but the status profiles provided must be periodically updated to ensure that they reflect actual quantities, usage, and age. Our study experience indicates the following:

- Keeping the status profiles current is key to their continued utility to help inform Army TWV strategy decisions.
- Improvements in data-collection processes and mechanisms can facilitate periodic updates of the TWV status profiles.
- Research on the EUL concept can lead to more accurate computation of EUL estimates.
- Detailed analyses can inform a holistic Army TWV strategy.
- Further research focused on the Army's knowledge base can lead to methods and recommended modifications for determining lifetime conditions of the medium and heavy TWV fleets.

Finally, we recommend that future research focus on the development of techniques that will enable the Army to visualize the impacts of strategic options and the effects of programming decisions. Such techniques would allow the Army to make more informed decisions in

Figure S.2
Where the Army Is: A Five-Year Window of Opportunity

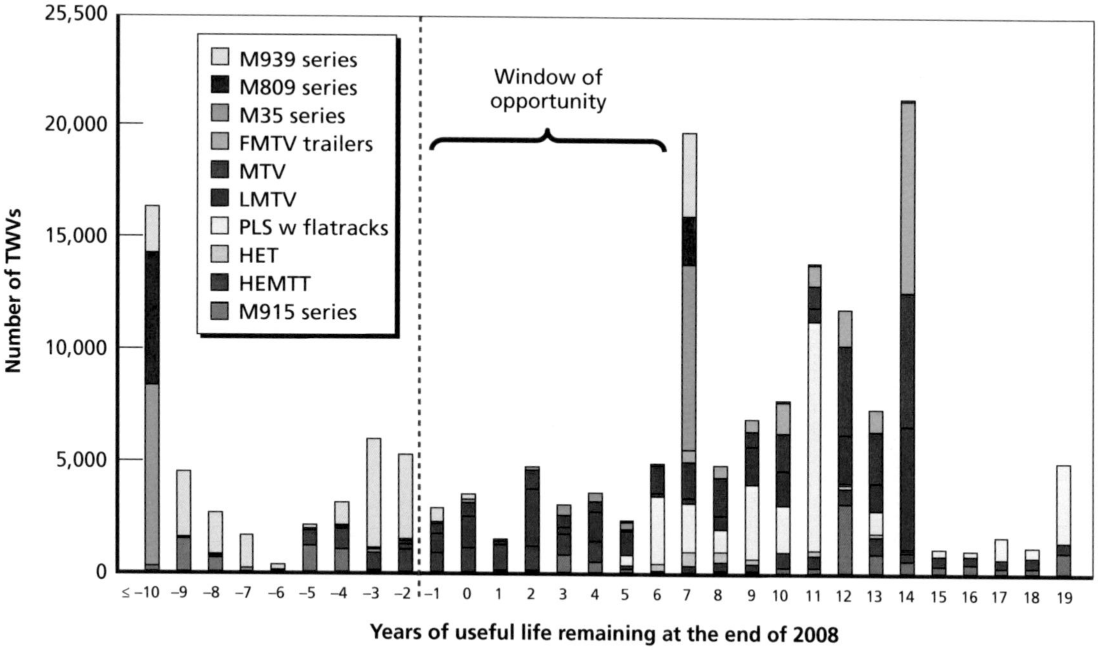

NOTE: Assumes 15-year EUL for medium TWVs and 20-year EUL for heavy TWVs.
RAND *TR890-S.2*

responding to programmatic changes as well as in designing TWV strategies that are effective and efficient in meeting the Army's future requirements.

Acknowledgments

The authors wish to thank BG Charles Anderson, former Director, Force Development, Deputy Chief of Staff, G-8, U.S. Army; MG David Halverson, Director, Force Development, Office of the Deputy Chief of Staff, G-8, U.S. Army; and Mr. Christopher Lowman, Deputy Chief of Staff, G-4, for sponsoring this research. In addition, the research team thanks LTC Robert Steigerwald, Mr. Kiyalan Batmanglidj, LTC Keith Rivers, Mr. Lee DeArmond, Dr. Cory Davis, and Ms. Felicia Walters for their able assistance as action officers during various periods of this study. The authors gratefully acknowledge the guidance and support provided by the LTV IPT and the TWV IPT. In particular, we thank Mr. Mike Zapf and Mr. Jose Rivera of the TWV IPT, both of whom provided data and valuable insights on heavy and medium tactical vehicles. Finally, the authors thank Mr. Thomas J. Edwards and Dr. Lisa Colabella for their thorough reviews and insightful comments.

Acronyms

A0	Model A0
A1	Model A1
A2	Model A2
A2+	Models A2, A3, A4, etc.
AFM	Army Flow Model
AKO	Army Knowledge Online
AMSAA	Army Materiel Systems Analysis Agency
AWRDS	Army War Reserve Deployment System
BG	Brigadier General
C2	Command and control
CHU	Container handling unit
CROP	Container roll in/out platform
DCS	Deputy Chief of Staff
DD250 Form	Materiel inspection and receiving report
EUL	Expected useful life
FMTV	Family of Medium Tactical Vehicles
FY	Fiscal year
GTA	Grow the Army
HEMAT	Heavy Expanded Mobility Ammunition Trailer
HEMTT	Heavy Expanded Mobility Tactical Truck
HET	Heavy Equipment Transporter
HIMARS	High Mobility Artillery Rocket System
HMMWV	High Mobility Multipurpose Wheeled Vehicle
HQDA	Headquarters, Department of the Army
HTV	Heavy Tactical Vehicle
IED	Improvised explosive device
ILAP	Integrated Logistics Analysis Program

IPT	Integrated Product Team
JLTV	Joint Light Tactical Vehicle
LHS	Load Handling System
LIW	Logistics Information Warehouse
LMTV	Light Medium Tactical Vehicle
LTAS	Long Term Armoring Strategy
LTG	Lieutenant General
LTV	Light Tactical Vehicle
LWB	Long Wheel Bed
LWV	Long Wheel Vehicle
MG	Major General
MRAP	Mine Resistant Ambush Protected
MTV	Medium Tactical Vehicle
NIPRNET	Non-Classified Internet Protocol Router Network
ODCSLOG	Office of the Deputy Chief Of Staff, Logistics
ODCSOPS	Office of the Deputy Chief of Staff, Operations and Plans
OMA	Operations and Maintenance, Army
OPA	Other Procurement, Army
OSMIS	Operating and Support Management Information System
PBUSE	Property Book Unit Supply Enhanced
PIP	Product Improvement Program
PLS	Palletized Load System
PM	Program Manager
POM	Program Objective Memorandum
RSV	Resupply Vehicle
SIPRNET	Secure Internet Protocol Network
TACOM	Tank-Automotive and Armaments Command
TPER	Theater provided equipment refurbishment
TWV	Tactical Wheeled Vehicle
UAH	Up Armored HMMWV
WebLIDB	Web-based Logistics Integrated Database
WSR	Weapon System Review
W/MHE	With man handling equipment
W/W	With winch

CHAPTER ONE

Introduction

The Army's medium and heavy tactical wheeled vehicle (TWV) fleets (both active and reserve components) are critical to sustaining its global operations: these are the vehicles that move supplies and equipment to and around the battlespace. The Army has maintained a significant program and made major investments in its medium and heavy TWV fleets because they are such critical assets. Table 1.1 shows that since fiscal year (FY) 2005, the Army has invested over $1 billion per year to procure medium and heavy TWVs; converting those amounts to FY 2009 dollars shows that the Army has invested more than $16 billion over the last five years. Nevertheless, the Army's TWV program has not been able to keep up with the demands of its aging fleets. Today there are medium and heavy vehicles that are over 30 years old. Perhaps more important, the pace and requirements of current operations, particularly in Iraq and Afghanistan, and predeployment training are stressing these fleets even more. Both the actual and imposed aging of these vehicles suggest that the Army needs to update its TWV strategy, which is a key element toward managing the Army's TWV investments prudently. In order to effectively update the TWV strategy, the Army must be able to make informed decisions about its investments in replacing, upgrading, and using its TWV fleets. Informed decisions are dependent on a clear understanding of how many of what types of vehicles the Army currently has and some indication of their age, condition, and remaining useful life as well as what the Army's requirements are likely to be at points in the future.

Table 1.1
The Army Has Major Investments in TWVs

Fleet	FY 2005 $M	FY 2006 $M	FY 2007 $M	FY 2008 $M	FY 2009 $M (projected)	Total in FY 2009 $M
Medium TWV	1081.3	674.8	3090.0	2147.0	1017.5	8329.4
Heavy TWV	612.4	369.5	1569.6	3095.8	1978.6	7836.1
FY total	1693.7	1044.3	4659.6	5242.8	2996.1	16165.6

SOURCES: U.S. Army Procurement Programs, "FY 2007 Budget Estimates: Other Procurement, Army, Activity 1, Tactical and Support Vehicles," *Committee Staff Procurement Backup Book*, February 2006.

U.S. Army Procurement Programs, "FY 2008 Budget Estimates: Other Procurement, Army, Activity 1, Tactical and Support Vehicles," *Committee Staff Procurement Backup Book*, February 2007.

U.S. Army Procurement Programs, "FY 2009 Budget Estimates: Other Procurement, Army, Activity 1, Tactical and Support Vehicles," *Committee Staff Procurement Backup Book*, February 2008.

U.S. Army Procurement Programs, "FY 2010 Budget Estimates: Other Procurement, Army, Activity 1, Tactical and Support Vehicles," *Committee Staff Procurement Backup Book*, May 2009.

NOTE: Army Fiscal Year 2009 Other Procurement, Army inflation indices used for conversion to FY 2009 dollars in millions. Amounts include Active Army, Army National Guard, and Army Reserve. Amounts also include supplemental funding identified in source documents.

When this study began, the Army had multiple databases containing vehicle on-hand quantities, but the quantities were often inconsistent among the databases and the differences were difficult to reconcile. While some databases did contain manufacture dates and odometer readings to indicate age and condition, the data were not complete enough to ascertain the age or lifetime condition of the vehicles.[1] In short, the Army's databases could not be readily used to determine comprehensive status profiles that show on-hand quantities and remaining useful life of the medium and heavy tactical wheeled vehicle fleets. This study built status profiles of these fleets that contain this critical information.

Background

The last formal Army TWV strategy was developed and signed in the 2004–2005 time frame. Since then, a number of changes have occurred. For example, the payload requirements have changed. Force protection is receiving greater emphasis. Doctrinal changes, as a result of current operations and the development of new capabilities, also suggest re-examination of the strategy. In a related manner, structural changes in the Army, such as modularization, and a significant growth in Army end-strength also have impacts on the Army's medium and heavy TWV fleets. In addition, new technologies offer important advances in terms of protection, mobility, payload, fuel efficiency, automation, and many other areas. Finally, TWVs are now procured with supplemental funds as well as Program Objective Memorandum (POM) funding.

The Army has revisited its 2005 strategy annually and updated it to reflect some of these changes. However, the most recent updates have focused primarily on the light fleet. In fact, a short-term RAND Arroyo Center study, discussed in Appendix C, focused on the light TWV fleet. In that effort, status profiles were provided to the study team as a starting point (assumptions) for analysis aimed at informing light TWV investment decisions. Although resources for the short-term effort did not allow for verification of the status profiles, the realization that status profiles were critical to the analysis served as a catalyst for the current effort. This study complements the Army's earlier efforts by focusing on the medium and heavy fleets. Though the Army has strived to update its TWV strategy, these efforts have had to be conducted with an incomplete understanding of the current status of the medium and heavy TWV fleets. Moreover, the updating efforts had to work around the fact that the Army does not have a formal process for creating an aggregate status profile of its medium and heavy TWV fleets.

Purpose

Updating its TWV strategy will require the Army to make numerous informed decisions about its vehicle replacement, upgrading, and usage plans. The purpose of this project is to provide current TWV fleet profiles that can inform fleet management analysis and decisions.

---

[1] For example, we found that 65 percent of the Heavy Expanded Mobility Tactical Truck (HEMTT) data are missing year of manufacture and/or correct odometer readings in our Operating and Support Management Information System (OSMIS) extract for that data.

Approach

The study team integrated data from various sources to construct status profiles of the Army's medium and heavy TWV fleets. Army data bases were used whenever possible. Non-Army sources were used to augment and validate Army data to ensure robustness in the fleet profiles.

Organization

Chapter Two describes the status profile building process. Chapter Three presents the status profiles of the medium and heavy tactical wheeled vehicle fleets. Chapter Four presents recommendations and closing remarks. Appendix A shows the detailed status profiles of the 40 models of heavy tactical vehicles included in this study. Appendix B shows the detailed status profiles of the 117 types of medium tactical vehicles included in this study. Appendix C summarizes RAND Arroyo Center's short-term effort on light TWVs that help set the stage for the current study. Appendix D discusses the state of the Army's data with respect to generating status profiles of its TWV fleets

CHAPTER TWO

Building the Status Profiles of the Heavy and Medium TWV Fleets

Data Sources

Table 2.1 shows the most promising data sources identified and used for this study. While not explicitly accessed by this study, many of the sources in Table 2.1 are fed by and/or feed the Army's Property Book Unit Supply Enhanced (PBUSE) system.[1]

Quantities of Vehicles by Type

Quantity data were available in a variety of forms, ranging from serial numbers to gross aggregate quantities for all TWVs or heavy TWVs or medium TWVs. Some quantities were available per fielding date, per manufacturing schedule, by fiscal year, by calendar year, by location, and by other schemes. The study team chose to use calendar year because all test cases indicated that either fiscal year or calendar data were generally available, though multiple data sources had to be consulted. Conversion to calendar year was more straightforward than conversion to fiscal year because the fiscal year starts on a different date each year, whereas the calendar year always begins on January 1 and ends on December 31.

Quantity is also dependent on the fidelity of roll-up level. Test cases revealed that in some cases data were available at the individual model level, but only at more aggregate levels for other vehicles. The study team collected or derived quantity data by model whenever possible and by the lowest aggregate level when data were not available at the individual model level.

Fleet Usage and Age

Vehicle age was used as an indicator for condition. In this case, the inherent assumption would be that the older a vehicle in terms of age, the more likely it would be to need recapitalization or replacement. Age could be measured from year of manufacture, from fielding date, from fiscal year of formal Army acceptance of the vehicle, or from some other date. Calendar year of manufacture was chosen as the most straightforward base. Although calendar year of manufacture was sometimes sparsely available for some models of medium and heavy TWV, multiple sources contained this information, and the study team determined that it was feasible to consolidate and cross-check year-of-manufacture data to ascertain a year of manufacture

---

[1] More information on this data source is available at https://www.pmlis.lee.army.mil/tls/pbuse/pbuse.htm

Table 2.1
Data Sources

Database Name	Description
Army Flow Model (AFM)	The AFM is an HQDA knowledge management system that provides the Army staff with the capability to analyze and assess actual or notional policy decisions over time. The AFM's primary purpose it so provide an effective and efficient means to assess the feasibility, supportability, and affordability of current, programmed, and hypothetical HQDA initiatives, and their impact on force readiness over time. AFM consists of an integrated suite of predictive models that enables the Army Staff and Commands to rapidly assess the effect of force structure and policy changes across the spectrum of functional and program elements. AFM integrates data from its suite of models with Army standard data and provides this data to the Army Staff and Commands through an easy-to-use web-based system as part of Arm Knowledge Online's (AKO) Operational Community on both NIPRNET and SIPRNET. Source: John McKitrick, article in ARMY AL&T, January 1, 2002.
Army War Reserve Deployment System (AWRDS)	AWRDS is an automated information system capable of building and maintaining databases containing Army War Reserve stocks and equipment data. This information reflects how the U.S. Army War Reserve stocks are configured to support rapid military deployment. AWRDS also assists in the development of U.S. Army Battle Books for War Reserve sites that list specific force structures (supplies and equipment) and associated embarkation plans. AWRDS is able to retrieve information and provide total asset visibility into containers and multi-pack items, in real time, in the form of reports, listings, and datasets. AWRDS utilizes bar code technology to collect equipment data and track and maintain changes in cargo configurations. Source: http://awrds.leapquest.com/ by 2007 Stanley Associates, Inc.
Contractor and Program Manager (PM)	The PM manages acquisition programs of direct interest to the DoD and services. A PM conducts acquisition studies, economic analyses, and related activities to ensure timely, critical, and cost-effective decisions on fleet modernization and readiness, prudent investment strategies, new system acquisition and deployments, and the continued usage and/or upgrade of existing assets via cost-effective technology insertion activities.
DD250: Materiel Inspection and Receiving Report	The DD250 form is a report used to indicate the government's inspection and acceptance of equipment or data, as well as an invoice for payment and a packing list.
FY 2008 TWV Integrated Product Team (IPT)	This IPT was formulated to assist with this study. The IPT included members from G-4, G-8, and other stakeholder organizations.
Jane's Military Vehicles and Logistics website	This website provides historical and overview information on military vehicles.
Logistics Information Warehouse (LIW)	LIW is the Army's Integrated Corporate Logistics Data Warehouse. The LIW provides streamlined web access to a host of essential Army logistics tools, including Parts Tracker, ILAP, WebLIDB, Army RESET Management Tool, and the new LIW Business Intelligence (BI) dashboard tools.
Operating and Support Management Information System (OSMIS)	The OSMIS Relational Database presents annual Operating and Support (O&S) historical information for Army Materiel Systems. The OSMIS Relational Database contains information on Aviation Systems, Combat Systems, Artillery/Missile Systems, Tactical Systems, Engineer/Construction Systems, Communications/Electronics Systems, and Data Processing Systems. Data sources used in these reports are from the U.S. Army Logistics Support Activity, Army Materiel Command, Major Subordinate Commands, the Industrial Operations Command, ODCSOPS, and ODCSLOG. Source: http://www.asafm.army.mil/ceac/cr/overview.asp, September 28, 2009.
President's Budgets FY 1997–2009 Other Procurement, Army (OPA)	The OPA contains historical (actuals) as well as projected cost, quantity, and schedule data for Army systems.
Theater Provided Equipment Refurbishment (TPER) data	Tactical TPER managers at TACOM Integrated Logistics Support Center provide data on repair and refurbishment of military equipment in theater.

for the vehicle groups included in the study.[2] As explained below, the year of manufacture was used as a starting point for a computation of years of useful life remaining, which the study adopted as a surrogate indicator of condition.

Expected Useful Life

For the purposes of this study, the expected useful life (EUL) of a vehicle is the time until it reaches a point of such extensive and widespread wear that it is more economical to recapitalize or replace the vehicle than to continue to maintain and repair it. We obtained estimates of peacetime EUL for each category of TWV. For light TWVs, the EUL estimate is 15 years. For medium vehicles, the EUL estimate is also 15 years, and for heavy TWVs, the EUL estimate is 20 years.[3] The EUL estimates we used are assumed values, not statistically derived values computed from empirical data. As such, it is important to note that the status profiles in this document can change if empirically derived EUL estimates differ from the assumed values.[4]

Combining the EUL estimates with age data allows us to compute the years of useful life remaining on groups of vehicles, and with that knowledge we can determine which groups of vehicles are likely to need immediate, near-term, mid-term, or far-term attention in terms of replacement or recapitalization. The years of useful life remaining can be calculated using the year of manufacture as the starting point to measure 20 years of EUL for heavy vehicles and 15 years of EUL for medium and light vehicles. Such calculations would allow vehicles with different EULs and different years of manufacture to be compared with respect to years of useful life remaining at a specific point in time. For example, a group of vehicles manufactured in 2000 with a EUL of 20 years will have 20 years – 8 years = 12 years of EUL remaining at the end of 2008. Similarly, a group of vehicles manufactured in 2001 with a EUL of 15 years will have 8 years of useful life remaining in 2008. In this hypothetical comparison, the average vehicle in the group of vehicles manufactured in 2001 with a EUL of 15 years is more likely to need recapitalization or replacement sooner than the average vehicle of the group of vehicles manufactured a year earlier in 2000 with a EUL of 20 years.

Status Data Base Building Process

In addition to established data sources, the project convened an Integrated Product Team (IPT) to assist with the study. The IPT had members from the study sponsor, G-4; the study coordinating sponsor, G-8; and members from other TWV stakeholder organizations. This IPT provided the study team with data, insights, and a variety of perspectives on TWV issues.

---

[2]   In cases where no dates were directly available, imputation based on existing evidence was used to assign a year of manufacture. For example, the serial number of a vehicle without a manufacture year was compared to serial numbers for vehicles with dates, and a manufacture year was assigned based on the proximity of serial numbers.

[3]   The expected useful life concept is used by the Army and other government agencies. For example, the Defense Logistics Agency estimates EUL values for various types of equipment as found in Circular A-076 issued by the Office of Management and Budget. Source: http://www.whitehouse.gov/omb/rewrite/circulars/a076/a076sa3.html

[4]   The Army Materiel Systems Analysis Agency (AMSAA) has been working with Tank-Automotive and Armaments Command (TACOM) in using statistical analysis techniques to try to determine true EUL values based on empirical data. Such values were not available for our study.

Converting the IPT-supplied data to a common format resulted in a base for the status data base. As might be expected, the resulting base had some gaps and some discrepancies. The study team used data from data sources shown in Table 2.1 to fill the gaps. For example, data were downloaded from the Logistics Information Warehouse (LIW) and OSMIS, and data were extracted from the President's Budgets. The extracted data were converted to the common format and overlaid onto the IPT base, filling gaps and creating overlaps. The overlaps were exploited to resolve discrepancies. When the overlaps were not adequate to resolve discrepancies, additional sources, such as Jane's Military Vehicles and Logistics websites, were consulted.

The HEMTT Test Case

The data base building process described above was refined through an application to a test case with Heavy Expanded Mobility Tactical Truck (HEMTT) vehicles.

The TWV IPT supplied the study team with two primary sets of data that pertained to HEMTT production and recapitalization activities. These sources were the Heavy Tactical Vehicles Weapon System Review (WSR), Volume 2, dated November 15, 2007, and the HEMTT Recapitalization Analysis Workbook. Data from these two IPT-supplied sources were converted to a common format to form the base of the HEMTT status data base. The study team generated data from the LIW and from the appendices of the President's Budgets for FY 1996–1997 through FY 2009. These efforts focused on filling data gaps and exploiting overlaps to resolve inconsistencies. The study team was able to generate two top-level recapitalization profiles by type and year. One was primarily based on the IPT-supplied data, and one was RAND-generated using all available sources. The RAND-generated historical profile was used to produce a HEMTT fleet years of useful life profile because it was more complete. Figure 2.1 shows the HEMTT recapitalization profiles generated.

The data sources show that the oldest HEMTT vehicles were originally manufactured in the early 1980s. Many of the older vehicles have since been recapitalized. Our goal was to combine the production and recapitalization data from all sources to produce a comprehensive historical profile of the HEMTT fleet that could be used to compute years of useful life.

The study team used vehicle serial numbers to compare production and recapitalization data on nearly 16,000 individual HEMTT vehicles. If a vehicle had been recapitalized, years of useful life remaining was computed from the recapitalization date. If a vehicle had not been recapitalized, years of useful life remaining was calculated from the original manufacture date. Recapitalized HEMTT vehicles were assumed to have a EUL of 20 years. Manufacturing and recapitalization are activities that span periods of time, and both are followed by fielding activities that span an additional period of time. To avoid the various interpretations of "manufacture date" or "recapitalization date" that could be imputed from the various sources of data (e.g., a donor vehicle received for recapitalization is associated with a particular date, a recapitalized vehicle ready for fielding has a later date, etc.), we standardized on using the date shown on the Army vehicle acceptance form—the DD250—for both the recapitalization date and the original manufacture date when such dates were available. In cases where DD250 dates were not available, imputation was used to determine a year of manufacture based on available

Figure 2.1
HEMTT Recapitalization Profiles

Model	FY97	FY98	FY99	FY00	FY01	FY02	FY03	FY04	FY05	FY06	Sum thru FY06	FY07	Total recap with date	Total recap w/ no date	Total recap	
M1120	0	0	3	107	45	3	1	3	3	3	156	3	156	2	158	RAND
M977	0	218	240	38	150	119	108	158	160	279	1,470	175	1,645	255	1,900	generated
M978	0	68	20	7	21	0	42	91	72	129	450	172	622	172	794	using all
M983	0	0	0	0	0	24	0	18	0	1	43	6	49	62	111	sources
M984	0	10	1	1	16	4	4	85	32	45	198	18	216	42	258	
M985	23	113	27	44	45	64	56	74	29	26	501	119	620	98	718	
Total	23	409	291	197	277	211	211	426	293	480	2,818	490	3,308	631	3,939	

									2,386		
M1120	31	126	112	141	209	114	149	251	1,133		
M977	0	0	0	0	23	0	185	5	213		
M978	0	0	0	0	174	0	143	90	407	HTV WSR Vol. 2	
M983	0	0	0	0	18	7	37	38	100	November 2007	
M984	0	0	0	0	116	0	7	16	139		
M985	0	0	0	21	90	6	28	0	145		
Total	31	126	112	162	630	127	549	400	2,137		

RAND *TR890-2.1*

evidence, e.g., continuity of serial numbers.[5] In some recapitalization cases, the donor vehicle and the resulting recapitalized vehicle were different models. The model of the recapitalized vehicle was used in our analysis, and the donor vehicle model was not counted.

Figure 2.1 shows that the WSR-generated profile counts a total recapitalized HEMTT fleet of 2,137 vehicles through FY 2006. Using all sources, the RAND-generated data shows a total recapitalized HEMTT fleet of 2,818 vehicles through FY 2006. The RAND-generated total through FY 2007 is 3,308 HEMTT vehicles. In addition, there are 631 vehicles that appear in at least one recapitalization data source. No DD250 recapitalization dates could be determined for these 631 vehicles, even though each could be associated with evidence that the vehicle existed (e.g., serial number with recapitalized model identified, listed on LIW download, etc.). If these 631 vehicles are included in the HEMTT fleet, then there were 3,939 HEMTT vehicles as of FY 2007. Evidence-based imputation was used to assign manufacture years to these 631 vehicles.

The recapitalization data shown in Figure 2.1 can be used to generate a historical profile of all HEMTT vehicles. Historical profiles can be used to ascertain the status of vehicle fleets. Figure 2.2 presents a HEMTT profile generated by combining the RAND-generated data of 3,939 recapitalized HEMTT vehicles with analogous data on HEMTTs that had not yet been recapitalized. The HEMTT fleet profile shows that over a third (34 percent) of the HEMTT fleet will exceed the IPT-supplied useful life estimate of 20 years by the end of FY 2008. More than another quarter (27 percent) will exceed useful life in the next five years, rendering a total

---

[5] For example, in the case where a vehicle had a serial number but no DD250 date, two reference vehicles with serial numbers consecutively before and after the vehicle missing a date were located and a manufacture year between the two reference dates was assigned to the vehicle with the missing DD250 date.

Figure 2.2
HEMTT Fleet Profile

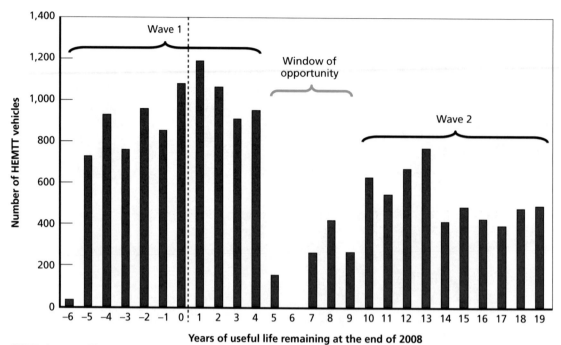

NOTE: Assumes 20-year EUL.

RAND *TR890-2.2*

of 61 percent of the HEMTT fleet beyond useful by 2013 if no additional recapitalizations are completed in 2008–2013.[6] In addition, unless HEMTT vehicles can be recapitalized more than once, a large number of replacements will have to be procured beginning in about 2017 to maintain Army requirements.

A fleet status profile can inform Army tactical wheeled vehicle strategy. The HEMTT fleet profile can be partitioned into three eras. The Army is now in the middle of the first era, a ten-year wave in which the quantity of HEMTTs exceeding EUL is rising rapidly. The second era is a three- to five-year window of opportunity from 2013–2018 when the rate of increase in the quantity of HEMTTs exceeding EUL slows. The third era is a second wave that arrives in about ten years, when the quantity of HEMTTs exceeding EUL will again rise rapidly. Strategies that deftly exploit the approaching window of opportunity may be able to minimize costs while restoring and maintaining a viable fleet of HEMTTs that meets Army requirements. For example, the Army can use the window of opportunity to recapitalize and/or replace vehicles in the first wave and then maintain a relatively steady rate of recapitalization/replacement for vehicles in the second wave. Such a strategy will likely be less costly than allowing the window of opportunity to slip by and then attempting to replace or recapitalize vehicles from both waves simultaneously. The cost benefits of exploiting the window of opportunity would arise because as the vehicles in the first wave grow older, their maintenance costs are likely to increase beyond the cost of replacing them. The earlier the vehicles in the first wave are replaced, the less costly it will be for the Army to maintain medium and heavy TWV capabilities.

---

[6] Analogous results can be generated for the 3,308 HEMTTs with complete recapitalization data.

CHAPTER THREE

Status Profiles of the Heavy and Medium TWV Fleets

The status profiles of the tactical wheeled vehicle fleets are shown in this chapter as years of useful life profiles. In each of the profile graphs, the X-axis shows the number of years of useful life remaining at the end of 2008. The Y-axis shows the number of vehicles. The different types of vehicles within each family of heavy vehicles are represented by the different colored bars on the graphs. The vertical red dotted line near the center of each chart marks the end of 2008. The vehicles represented by the bars to the left of the dotted line have exceeded useful life as of the end of 2008. The vehicles represented by the bars to the right of the red dotted line have years of useful life remaining as of the end of 2008.

Table 3.1 shows all of the heavy tactical wheeled vehicles included in this study and the categorization of the vehicles assumed for this study. Table 3.2 shows all of the medium tactical wheeled vehicles included in this study and the categorization of the vehicles assumed for this study.

Appendix A shows the useful life profiles of the 40 individual models of heavy TWVs. Appendix B shows the useful life profiles of the 117 individual models of medium TWVs.

Table 3.1
Heavy Tactical Wheeled Vehicles

Family	Type/Model
HEMTT	M978 Tanker M983 Tractor M984 Wrecker M977 Cargo M985 Cargo M989 Heavy Expanded Mobility Ammunition Trailer (HEMAT) M1120 Load Handling System (LHS)
HET	M1070 M1000 Semitrailer
PLS	M1074 Truck M1075 Truck M1076 PLS Trailer (16.5 ton) Container Handling Unit (CHU) M1077 Container Roll-In/Out Platform (CROP) M1 Flatrack
M915 series	M915 Line Haul • M915 • M915-1207 • M915A1 • M915A2 • M915A2P1 • M915A2P1-2482 • M915A3-1942 • M915A3-4847 • M915A3P1 • M915A4P1 • M919 • RC2664T-3616 • RC2664T-8623 • RC2664T-8624 M916 Light Equipment Transporter (LET) • M916 • M916A1 • M916A1P1 • M916A2 • M916A3P1 M917 • M917 • M917-1165 • M917-4389 • M917-6963 • M917-8249 M920 Transporter

Table 3.2
Medium Tactical Wheeled Vehicles

Family of Medium Vehicles						
LMTV	MTV	Trailers	M35 Series	M809 Series	M939 Series	
M1078 Cargo	M1083 Cargo	M1082 LMTV	M35-2046	M809-9007	M923	M930A2
M1078 Cargo W/W	M1083 Cargo W/W	XM1147 LHST	M35-2047	M810-0586	M923A1	M931
M1081 Air Drop Cargo	M1093 Air Drop Cargo	M1095 MTV	M35-2048	M813-8890	M923A1P1	M931A1
M1081 Air Drop Cargo W/W	M1093 Air Drop Cargo W/W	M1095/RST	M35-2049	M813-8902	M923A2	M931A1P1
M1079 Van	M1085 LWB Cargo		M35-3850	M813A1-8905	M923A2P1	M931A2
M1079 Van W/W	M1085 LWB Cargo W/W		M35-6568	M813A1-8913	M924	M931A2P1
M1080 Chassis	M1084 Cargo W/MHE		M35-8463	M814-8987	M925	M932
	M1086 LWB Cargo W/MHE		M35-8464	M814-8988	M925A1	M932A1
	M1090 Dump		M35A1-5633	M815	M925A1P1	M932A1P1
	M1090 Dump W/W		M35A1-5634	M816	M925A2	M932A2
	M1094 Air Drop Dump		M35A2-1616	M817-0589	M925A2P1	M932A2P1
	M1094 Air Drop Dump W/W		M35A2-1617	M817-8970	M927	M934
	XM1157 10-Ton Dump		M35A2C-0873	M818-8978	M927A1	M934A1
	XM1157 10-Ton Dump W/W		M35A2C-0875	M818-8984	M927A2	M934A2
	M1088 Tractor		M35A2C-2050	M819	M927A2P1	M935
	M1088 Tractor W/W		M35A2-1618	M820	M928	M935A1
	M1089 Wrecker W/W		M35A2-1619	M820A2	M928A1	M936
	M1087 Exp Van			M821	M928A1P1	M936A1
	XM1148 LHS				M928A2	M936A1P1
	M1092 Chassis				M929	M936A2
	M1096 LWB Chassis				M929A1	M936A2P1
	M1084/RSV HIMARS RSV				M929A1P1	M942A2-0287
					M929A2	M942A2-0289
					M929A2P1	M944A2
					M930	

Heavy Tactical Wheeled Vehicles

Four families of heavy tactical wheeled vehicles were included in this study: the Heavy Expanded Mobility Tactical Truck (HEMTT), the Heavy Equipment Transporter system (HET), the Palletized Loader System (PLS), and the M915 series of heavy vehicles.

Heavy Expanded Mobility Tactical Truck

The years of useful life profile for the HEMTT family of vehicles is discussed in Chapter Two and shown in Figure 2.2.

Heavy Equipment Transporter

The years of useful life profile for the HET family of vehicles is shown in Figure 3.1. As can be seen, the entire HET fleet has at least six years of useful life remaining as of the end of 2008.

Palletized Load System

The years of useful life profile for the PLS family of vehicles is shown in Figure 3.2. As illustrated, the entire PLS fleet has at least five years of useful life remaining as of the end of 2008.

M915 Series of Heavy Tactical Vehicles

The years of useful life profile for the M915 family of heavy vehicles is shown in Figure 3.3. As can be seen, 35 percent of the M915 fleet exceeded expected useful life at the end of 2008, and 70 percent will exceed expected useful life by the end of 2020.

Figure 3.1
HET Fleet Profile

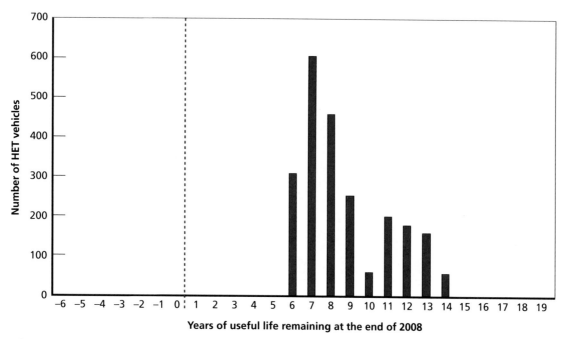

NOTE: Assumes 20-year EUL.
RAND *TR890-3.1*

Figure 3.2
PLS Fleet Profile

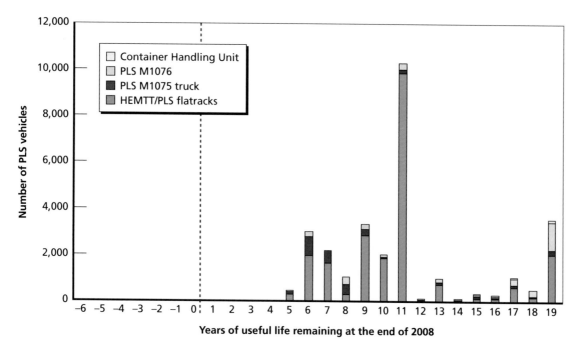

NOTE: Assumes 20-year EUL.
RAND *TR890-3.2*

Figure 3.3
M915 Series Fleet Profile

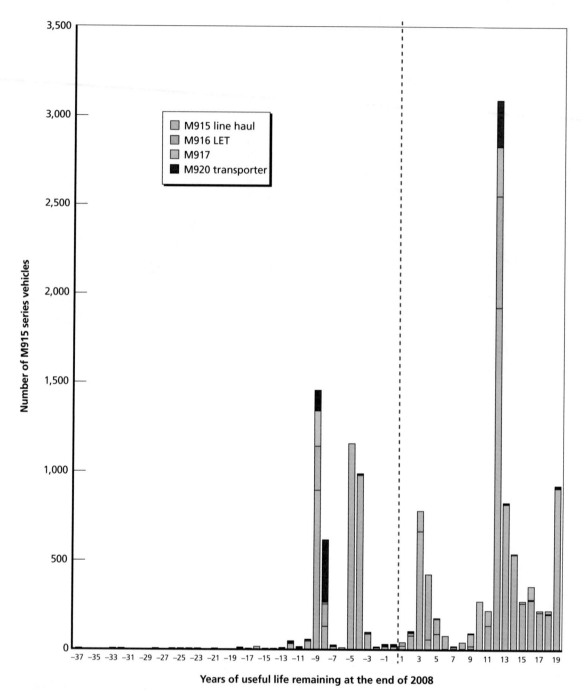

NOTE: Assumes 20-year EUL.

RAND *TR890-3.3*

Aggregate Heavy Tactical Wheeled Vehicle Fleet Profile

Figure 3.4 shows the aggregate heavy TWV fleet years of useful life profile with PLS flatracks. Figure 3.5 shows the aggregate heavy TWV fleet years of useful life profile without PLS flatracks.

One observation that stands out in the graphs shown in Figure 3.4 and Figure 3.5 is that the only vehicles to have exceeded useful life as of the end of 2008 are M915 series trucks and the older HEMTT vehicles. All of the PLS and HET vehicles have years of useful life remaining as of the end of 2008. Hence, an Army strategy for heavy TWVs based on these graphs should take into account that a substantial portion of the M915 and older HEMTT vehicles have already exceeded useful life and therefore are likely to be the heavy TWVs most in need of recapitalization or replacement.

Figure 3.4
Aggregate Heavy Tactical Wheeled Vehicle Fleet Profile With PLS Flatracks

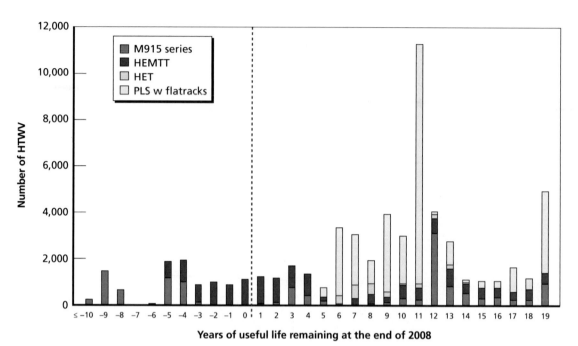

NOTE: Assumes 20-year EUL.
RAND *TR890-3.4*

Figure 3.5
Aggregate Heavy Tactical Wheeled Vehicle Fleet Profile Without PLS Flatracks

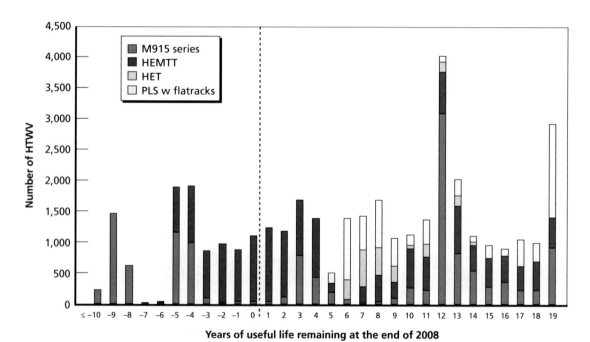

NOTE: Assumes 20-year EUL.
RAND *TR890-3.5*

Medium Tactical Wheeled Vehicles

Four families of medium TWVs were included in this study. First, there is the Family of Medium Tactical Vehicles (FMTV), of which there are three major types: the Light Medium Tactical Vehicles (LMTV), the Medium Tactical Vehicles (MTV), and the medium trailers. Second, there is the M35 series of 2.5-ton medium trucks. Third, there is the M809 series of 5-ton medium trucks. Finally, there is the M939 series of 5-ton medium trucks. The aggregate medium TWV fleet graph (Figure 3.12) shows all four families of medium TWVs together.

Light Medium Tactical Vehicle

The years of useful life profile for the LMTV is shown in Figure 3.6. The LMTV profile is dominated by LMTV cargo trucks. Thirty-six percent of the LMTV fleet will exceed useful life within the next five years. Fifty-six percent of the LMTV fleet will exceed useful life within the next ten years. As of the end of 2008, the LMTVs were beginning to exceed useful life.

Medium Tactical Vehicle

Figure 3.7 shows the years of useful life profile for the MTV fleet. The MTV fleet profile is dominated by MTV cargo trucks. Fourteen percent of the MTV fleet will exceed useful life in the next five years, so there is only a small current replacement need. There will be a gradual increase in the quantity of MTVs exceeding useful life during the next fourteen years. A small number of MTVs exceeded useful life at the end of 2008.

Family of Medium Tactical Vehicle Trailers

Figure 3.8 shows the years of useful life profile of the FMTV trailer fleet. Because this fleet is relatively new, all the FMTV trailers have at least seven years of useful life remaining as of the end of 2008.

Figure 3.6
LMTV Fleet Profile

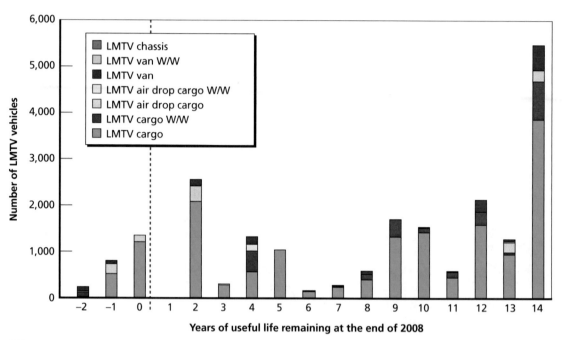

NOTE: Assumes 15-year EUL.
RAND *TR890-3.6*

Figure 3.7
MTV Fleet Profile

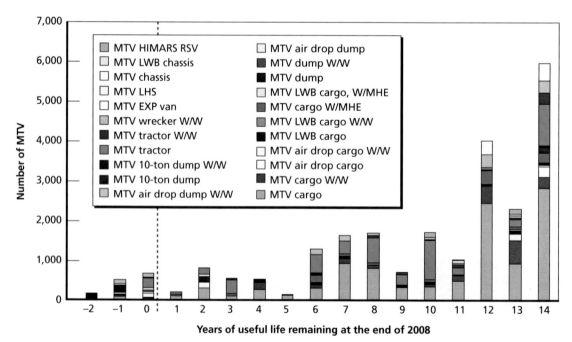

NOTE: Assumes 15-year EUL.
RAND *TR890-3.7*

Figure 3.8
FMTV Trailer Fleet Profile

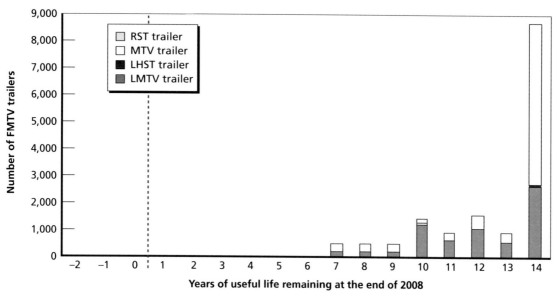

NOTE: Assumes 15-year EUL.
RAND *TR890-3.8*

M35 Series 2.5-Ton Medium Trucks

Figure 3.9 shows the years of useful life profile for the M35 series of 2.5-ton medium trucks. Forty-eight percent of the M35 fleet exceeded useful life at the end of 2008. An additional 7 percent will exceed useful life in the next six years. However, by 2015, 99 percent of the M35 fleet will have exceeded useful life. The Army has made a decision to divest the M35 series 2.5-ton medium trucks but has not yet implemented a plan to remove these trucks from service.

Although the available data results in the graph shown in Figure 3.9, we note a possible anomaly at the seven-year mark and recommend that future efforts include an investigation into the data for that particular year. The Army should not make TWV strategy decisions based on the year 7 data shown in Figure 3.9 without thoroughly re-examining all data sources and any anecdotal evidence that can be uncovered.

M809 Series 5-Ton Medium Trucks

Figure 3.10 shows the years of useful life profile for the M809 series 5-ton medium trucks. Seventy-four percent of the M809 series fleet exceeded useful life at the end of 2008. Within the next seven years, 99 percent of the M809 fleet will have exceeded useful life. The Army has made a decision to divest the M809 series 2.5-ton medium trucks but has not yet implemented a plan to remove these trucks from service. Although the data available results in the graph shown in Figure 3.10, we note a possible anomaly at the seven-year mark and recommend that future efforts include an investigation into the data for that particular year. The Army should not make TWV strategy decisions based on the year 7 data shown in Figure 3.10 without thoroughly re-examining all data sources and any anecdotal evidence that can be uncovered.

Figure 3.9
M35 Series 2.5-Ton Truck Fleet Profile

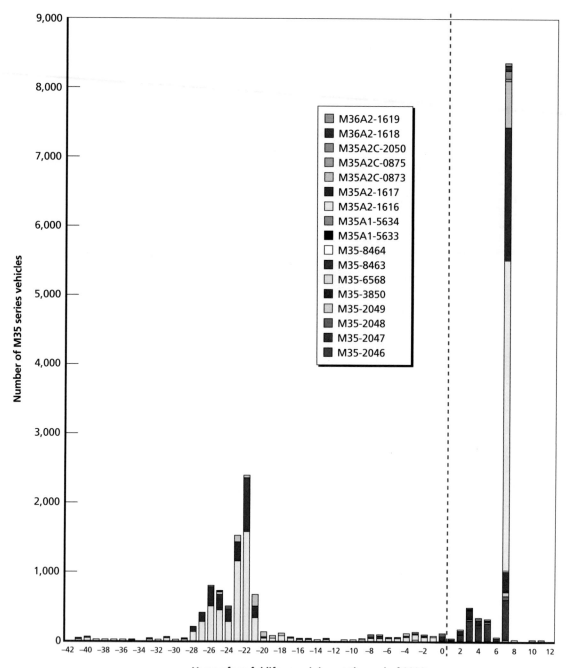

NOTE: Assumes 15-year EUL.

Figure 3.10
M809 Series 5-Ton Truck Fleet Profile

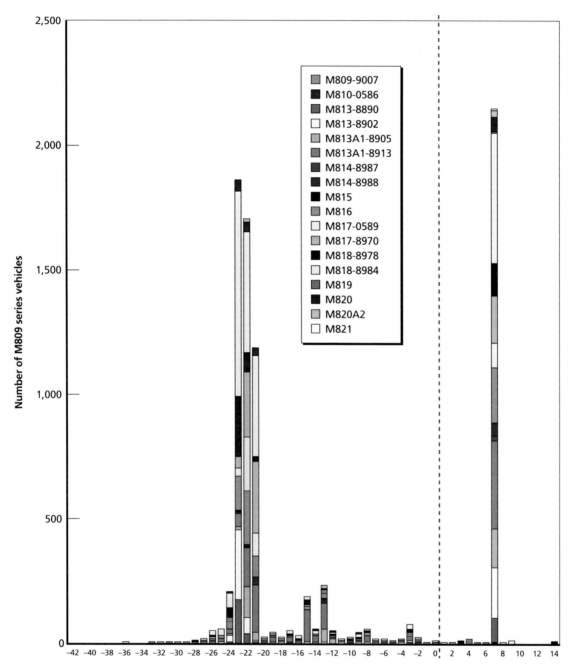

NOTE: Assumes 15-year EUL.

M939 Series 5-Ton Medium Trucks

Figure 3.11 shows the years of useful life profile for the M939 series 5-ton medium trucks. This family is called the M939 series even though it includes the M923 model. Eighty-two percent of the M939 fleet exceeded useful life at the end of 2008. By the end of 2015, 99 percent of the M939 fleet will have exceeded useful life. Although the status profile of the M939 fleet is quite similar to those of the M809 fleet and the M35 fleet, the Army is not planning to divest the M939 fleet. M939 vehicles, then, are among the oldest of the Army's TWVs destined to remain in service. Although the data available result in the graph shown in Figure 3.11, we note a possible anomaly at the seven-year mark and recommend that future efforts include an investigation into the data for that particular year. The Army should not make TWV strategy decisions based on the year 7 data shown in Figure 3.11 without thoroughly re-examining all data sources and any anecdotal evidence that can be uncovered.

Aggregate Medium Tactical Wheeled Vehicle Fleet Profile

Figure 3.12 shows the aggregate medium TWV fleet years of useful life profile. This graph shows that three types of medium vehicles have exceeded useful life as of the end of 2008. These are the older M939 series vehicles, the M809 series vehicles, and the M35 series vehicles. Since the Army has already decided to divest the M809 series and M35 series vehicles, the medium vehicles of interest that have already exceeded useful life are the M939 series. The graph also shows that MTV and LMTV are beginning to exceed useful life. Based on this profile, a strategy for medium TWV should reflect that the groups of M939 series vehicles represented by the bars to the left of the red dotted line in Figure 3.12 are the most likely to need recapitalization or replacement. In addition, a medium TWV strategy should consider that recapitalization or replacement of MTVs and LMTVs will likely need to follow.

Figure 3.11
M939 Series 5-Ton Truck Fleet Profile

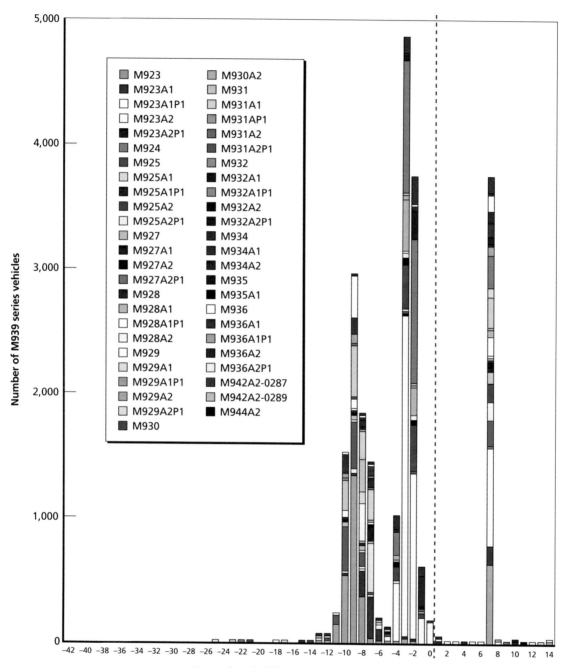

NOTE: Assumes 15-year EUL.

Figure 3.12
Aggregate Medium Tactical Wheeled Vehicle Fleet Profile

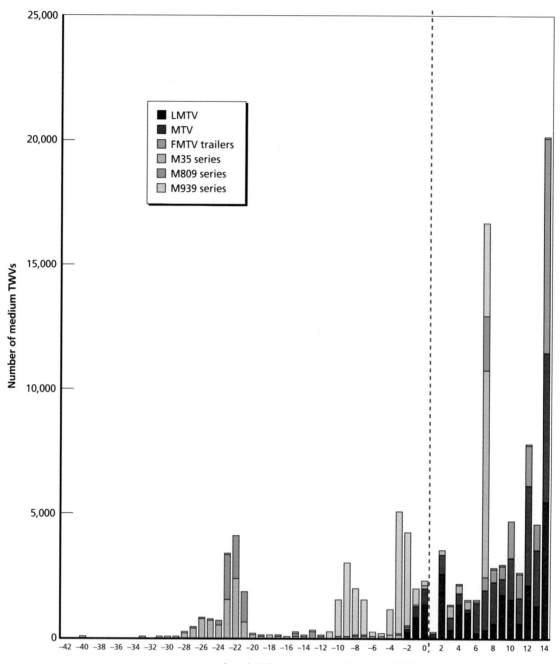

NOTE: Assumes 15-year EUL.
RAND *TR890-3.12*

Where the Army Is Now: The Army's Oldest Tactical Wheeled Vehicles

Figure 3.13 shows the Army's oldest tactical wheeled vehicles, combining the two aggregate graphs of heavy and medium vehicles shown in Figure 3.4 and Figure 3.12. The combination shows the years of useful life profile for the Army's TWVs that have exceeded their useful lives by ten years or more. It is evident that the Army has some very old vehicles. The oldest TWVs are the M939 series medium vehicles, the M809 series medium vehicles, and the M35 series medium vehicles. Since the Army has already decided to divest the M809 and M35, the vehicles of greatest interest on this graph are the M939 series, represented by the gold-colored bars. The Army has in inventory over two thousand M939 series medium vehicles that as a group have exceeded useful life by ten years or more. While an Army strategy must consider a variety of factors—including, but not limited to, usage, location, depreciation rates, maintenance costs, recapitalization costs, replacement costs, and military mission requirements—the graph in Figure 3.13 can help inform the decisionmaking process by indicating which vehicles have exceeded useful life and which still have years of useful life remaining.

Figure 3.13
The Army's Oldest Tactical Wheeled Vehicles

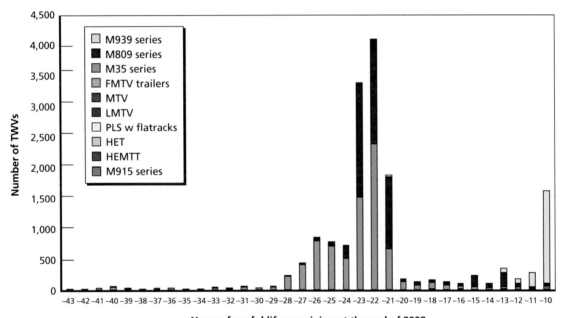

NOTE: Assumes 15-year EUL for medium TWVs and 20-year EUL for heavy TWVs.
RAND *TR890-3.13*

Where the Army Is Now: Five-Year Window of Opportunity

Figure 3.14 shows where the Army is now in terms of its medium and heavy TWV fleets. All the vehicles shown in Figure 3.13 are included in the left-most bar labeled "≤ –10" in Figure 3.14. In addition, Figure 3.14 combines the graphs shown in Figure 3.4 and Figure 3.12. This overview of the status of the medium and heavy TWV fleets shows that the Army is in the middle of a window of opportunity. That is, the Army has just entered a period where relatively few groups of medium and heavy TWVs are exceeding their useful life. An effective Army strategy would seek to exploit this window of opportunity to take care of those vehicles that have already exceeded expected useful life before the next wave hits in about five years. And although five years seems like a lengthy amount of time, the luxury is illusory. POM 2011 is near and POM 2012 will soon command the Army's attention, so from a planning point of view, the real window of opportunity is closer to two or three years. To take maximum advantage, the Army needs to begin immediate efforts to devise a TWV strategy that will serve it well into the future. The graph in Figure 3.14 indicates a likely starting point on which to base the Army's updated TWV strategy.

Figure 3.14
Where the Army Is Now: Five-Year Window of Opportunity

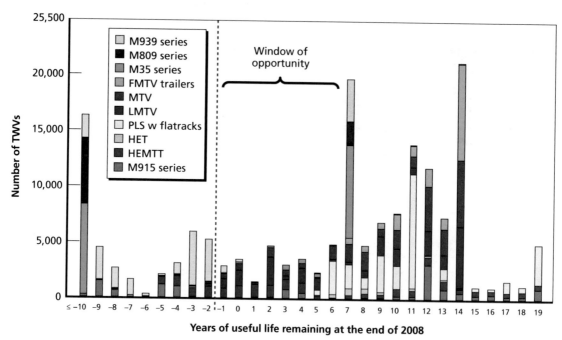

NOTE: Assumes 15-year EUL for medium TWVs and 20-year EUL for heavy TWVs.
RAND TR890-3.14

CHAPTER FOUR

Recommendations and Closing Remarks

The analysis in this study has produced status profiles of the medium and heavy TWV fleets. Such profiles can provide a starting point from which to build an effective Army TWV strategy. Other important factors to consider in formulating an Army TWV strategy include forecasts of TWV capability requirements, maintenance costs, cost to replace, cost to recapitalize, depreciation rates, and location. Further analysis that integrates the status profiles with the other factors will enhance the Army's ability to make informed replacement and recapitalization decisions regarding its medium and heavy TWV fleets. For example, one method of integrating the multiple factors is via a fleet management model. The status profiles along with other factors could be inputs to such models and allow the Army to evaluate the impact of TWV recapitalization and replacement decisions.

This study points to two areas that will enhance the continued utility of TWV status profiles that accurately reflect the actual quantities and condition of the Army's TWV fleets. The first area is the need to periodically update the status profiles. The quantities, usage, condition, and age of TWV fleets change with time. Hence, status profiles such as the ones generated in this study need to be kept current in order to continue to accurately reflect reality. The second area is the need to conduct detailed analysis of each individual model. The sections below discuss the two areas in more detail.

Maintaining TWV Status Data Bases

Readily accessible, accurate, consistent, and complete data are critical to updating the status profiles presented in this study. Our experience suggests that the Army has adequate frameworks and mechanisms for collecting TWV quantity and age data, but improvements in the collection process are needed to make the existing data bases useful for updating and providing status snapshots of the Army's medium and heavy TWV fleets. While this study did not examine the data-collection process, our experience suggests that stronger discipline, perhaps with the aid of automated data-collection mechanisms, higher priority and increased resources dedicated to the data-collection task, and recognition of the value of complete and consistent accessible data that accurately reflects reality, will all improve the process.

Further research is required to formulate methods to determine lifetime condition indicators using existing data bases. Such investigations will also be valuable for identifying modifications to existing data bases that could serve to make the data in the Army repositories useful in determining lifetime condition of TWVs. Such modifications could be included in future planned upgrades.

The Role of EUL

Currently, the EUL of medium and heavy vehicles is used by various Army agencies,[1] but little literature is available on its derivation or long-term accuracy. Research could uncover empirical methods for updating EUL estimates. For example, a mandate to accurately record reliable current data within the Army's existing data base frameworks could give the Army reliable and accurate usage records that could be used to compute EULs based on empirical data. Over time, such calculations can provide insights to deriving more accurate EUL estimates for future TWV models. Efforts such as the statistical analysis of empirical data being conducted by AMSAA and TACOM are steps in this direction.

Advances in Technology

In the future, technology can play a role in helping the Army maintain current status profiles of its TWV fleets. Automated data collection is one such technology. For example, universal identification systems exist and are used by various commercial industries to collect tracking and other data on individual items moving through a system. Such capabilities could enable automatic or semi-automatic data collection on TWVs for the Army. Considerable undertakings may be required to implement such a system. For example, the Army would have to decide which TWVs to track, what data to collect, how to collect the desired data, how to outfit the TWVs to allow information to be collected, when to collect the data, design the collection processes, identify the responsible parties, make cost/benefit determinations, and resolve a host of related issues to implement such a system. Gradual gravitation toward employing such technologies may be the most cost-effective method for realizing potential benefits, but such a transformation process has to be a planned integration into the TWV strategy.

The Need for Detailed Analysis

A "big picture" view of the TWV fleets can provide valuable insights on how best to formulate an effective Army TWV strategy. Overviews can identify the systems most in need of recapitalization or replacement attention. However, recapitalization and replacement needs can vary greatly by individual model. Detailed analysis can pinpoint the exact models that require attention. For example, the 186 MTV air drop dump trucks and the 178 MTV 10-ton dump trucks illustrate the value of detailed analysis. Figure 4.1 shows that all 186 air drop dump trucks will exceed their useful life by the end of 2010. Figure 4.2 shows that all 178 10-ton dump trucks will have at least ten years of useful life remaining at the end of 2010. Based on these two figures, the recapitalization and replacement needs of the MTV air drop dump trucks are likely to be required sooner than the recapitalization and replacement needs of the MTV 10-ton dump trucks.

More extensive detailed analysis could reveal that the air drop dump trucks have experienced very low usage because the need to perform the air drop function has been minimal

---

[1]   For an example, see BG John R. Bartley, PEO, CS & CSS, "NDIA 2008 Tactical Wheeled Vehicle (TWV) Conference," February 5, 2008, p. 21.

Figure 4.1
Air Drop Dump Truck Fleet Profile

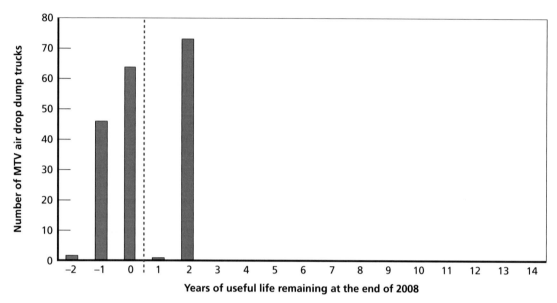

NOTE: Assumes 15-year EUL.
RAND *TR890-4.1*

Figure 4.2
MTV 10-Ton Dump Truck Fleet Profile

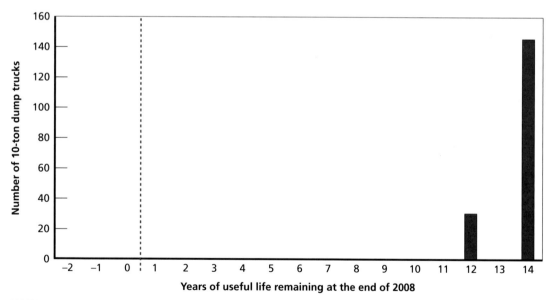

NOTE: Assumes 15-year EUL.
RAND *TR890-4.2*

while the 10-ton dump trucks have incurred very high usage. Actual usage, then, can dictate which model is in greater need for recapitalization or replacement. However, actual accurate and reliable usage data is often not available, so until such data is available for such detailed analysis, the Army should consider the EUL graphs supplied throughout this report in formulating its TWV strategy. The EUL graphs are a better indication of usage than no data.

Recommendations and Closing Remarks

If the Army is to effectively update its TWV strategy, it must know the current status of its TWV fleets in terms of quantities by type and some indication of overall condition. Without this information, the Army does not have a starting point to make decisions on how to meet its future TWV requirements; the Army must know where it is in terms of its current TWV fleets in order to determine how best to get to where it needs to be in the future.

This study has provided the Army with starting points for the medium and heavy TWV fleets, but the status profiles provided must be periodically updated to ensure that they reflect actual quantities, usage, and age. Our study experience indicates the following:

- Keeping the status profiles current is key to their continued utility to help inform Army TWV strategy decisions.
- Improvements in data-collection processes and mechanisms can facilitate periodic updates of the TWV status profiles.
- Research on the EUL concept can lead to more accurate computation of EUL estimates.
- Detailed analyses can inform a holistic Army TWV strategy.
- Further research focused on the Army's knowledge base can lead to methods and recommended modifications for determining life time conditions of the medium and heavy TWV fleets.

Finally, we recommend that future research focus on the development of techniques to enable the Army to visualize the impacts of strategic options and the effects of programming decisions. Such techniques would allow the Army to make more informed decisions in responding to programmatic changes as well as in designing TWV strategies that are effective and efficient in meeting the Army's future TWV requirements.

Profiles of the Individual Heavy TWV Models

M915 Line Haul Profile

Figure A.1
M915 Line Haul Fleet Profile

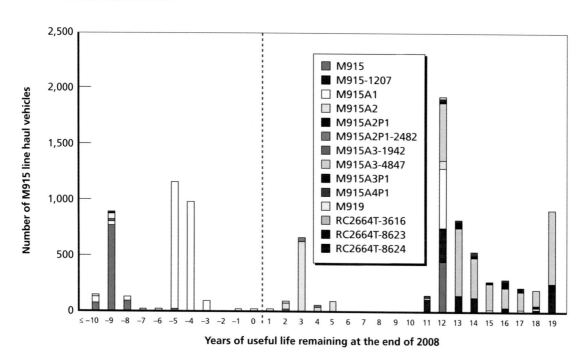

NOTE: Assumes 20-year EUL.
RAND *TR890-A.1*

RC2664T Fleet Profile

Figure A.2
RC2664T Fleet Profile

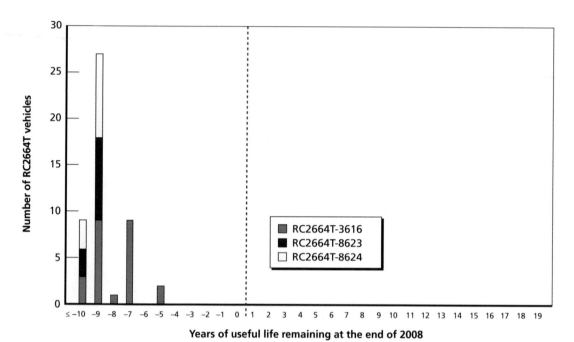

NOTE: Assumes 20-year EUL.

RAND *TR890-A.2*

M916 Series Fleet Profile

Figure A.3
M916 Series Fleet Profile

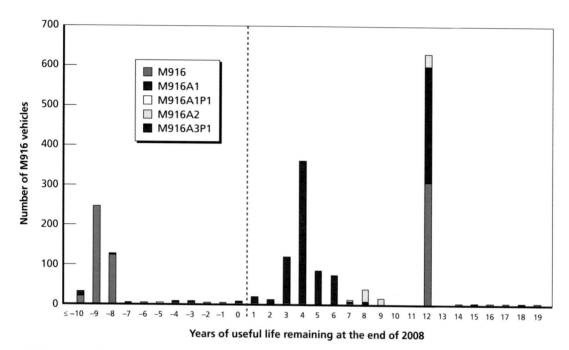

NOTE: Assumes 20-year EUL.
RAND *TR890-A.3*

M917 Series Fleet Profile

Figure A.4
M917 Series Fleet Profile

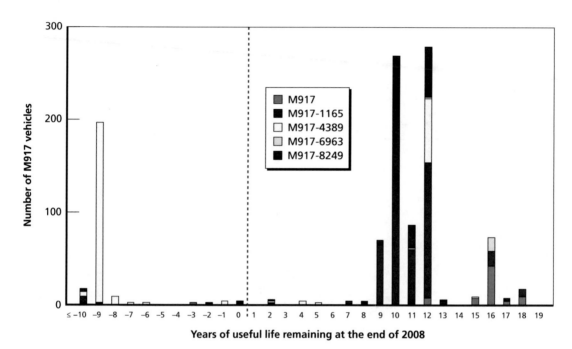

NOTE: Assumes 20-year EUL.
RAND *TR890-A.4*

M920 Transporter Fleet Profile

Figure A.5
M920 Transporter Fleet Profile

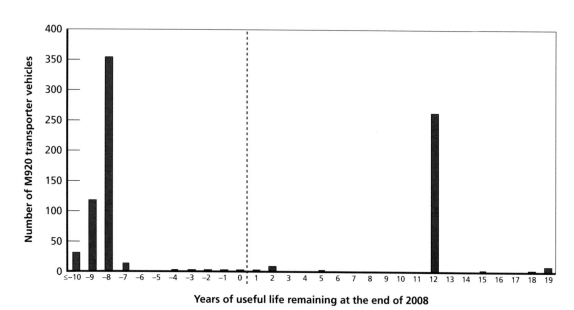

NOTE: Assumes 20-year EUL.

RAND *TR890-A.5*

HEMTT/PLS Flatrack Fleet Profile

Figure A.6
HEMTT/PLS Flatrack Fleet Profile

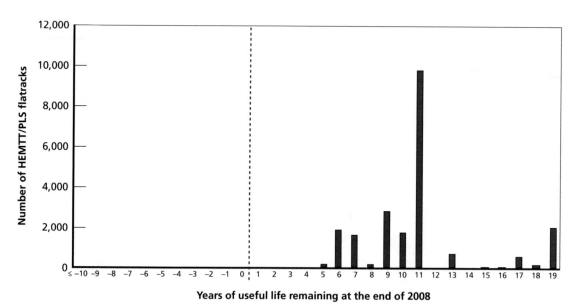

NOTE: Assumes 20-year EUL.
RAND *TR890-A.6*

PLS M1075 Truck Fleet Profile

Figure A.7
PLS M1075 Truck Fleet Profile

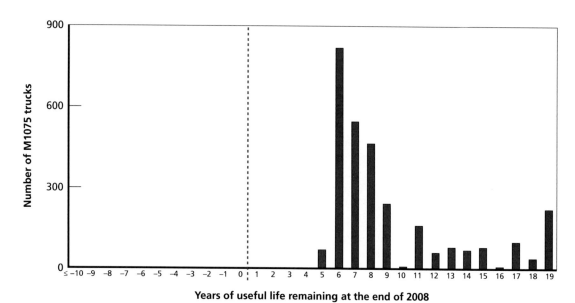

NOTE: Assumes 20-year EUL.
RAND *TR890-A.7*

PLS M1076 Trailer Fleet Profile

Figure A.8
PLS M1076 Trailer Fleet Profile

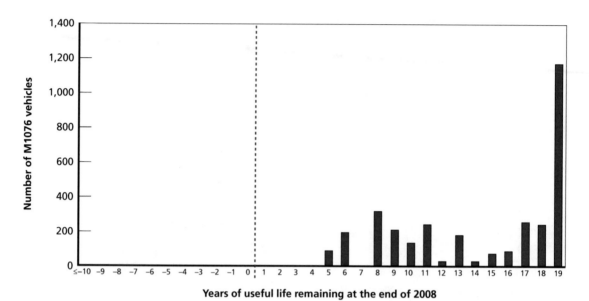

NOTE: Assumes 20-year EUL.
RAND *TR890-A.8*

Container Handling Unit Fleet Profile

Figure A.9
Container Handling Unit Fleet Profile

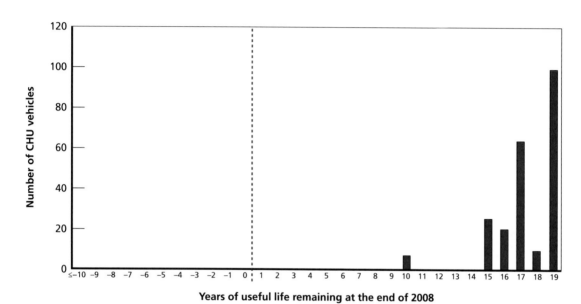

NOTE: Assumes 20-year EUL.
RAND *TR890-A.9*

Profiles of the Individual Medium TWV Models

LMTV Cargo

Figure B.1
LMTV Cargo Fleet Profile

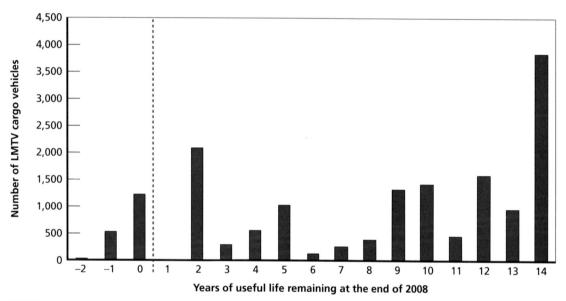

NOTE: Assumes 15-year EUL.
RAND *TR890-B.1*

LMTV Cargo with Winch Fleet Profile

Figure B.2
LMTV Cargo with Winch Fleet Profile

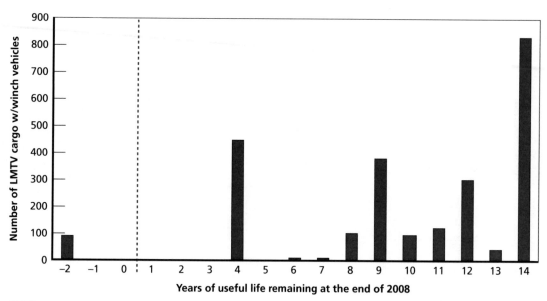

NOTE: Assumes 15-year EUL.
RAND *TR890-B.2*

LMTV Air Drop Cargo Fleet Profile

Figure B.3
LMTV Air Drop Cargo Fleet Profile

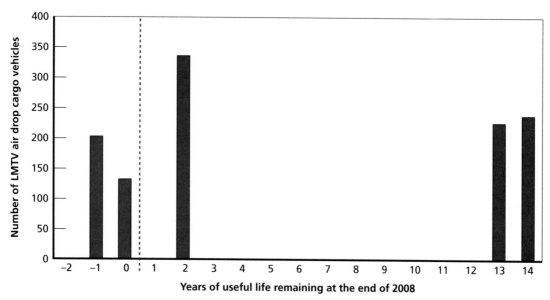

NOTE: Assumes 15-year EUL.
RAND *TR890-B.3*

LMTV Air Drop Cargo with Winch Fleet Profile

Figure B.4
LMTV Air Drop Cargo with Winch Fleet Profile

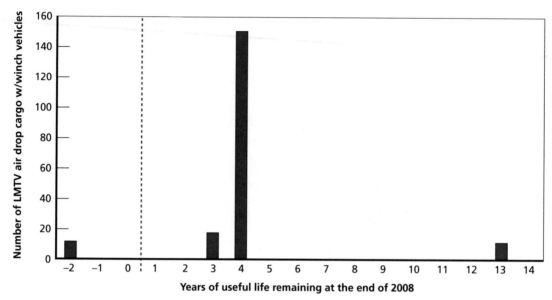

NOTE: Assumes 15-year EUL.
RAND *TR890-B.4*

LMTV Van Fleet Profile

Figure B.5
LMTV Van Fleet Profile

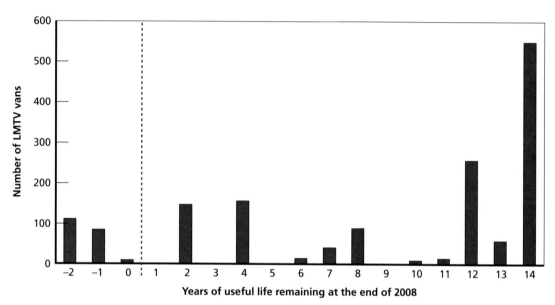

NOTE: Assumes 15-year EULs.
RAND *TR890-B.5*

LMTV Chassis Fleet Profile

Figure B.6
LMTV Chassis Fleet Profile

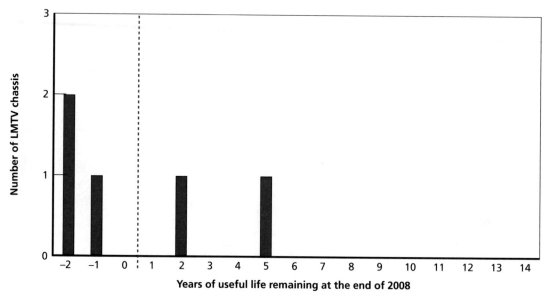

NOTE: Assumes 15-year EUL.
RAND *TR890-B.6*

MTV Cargo Fleet Profile

Figure B.7
MTV Cargo Fleet Profile

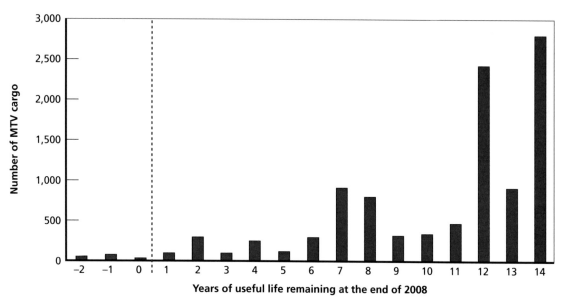

NOTE: Assumes 15-year EUL.
RAND *TR890-B.7*

MTV Cargo with Winch Fleet Profile

Figure B.8
MTV Cargo with Winch Fleet Profile

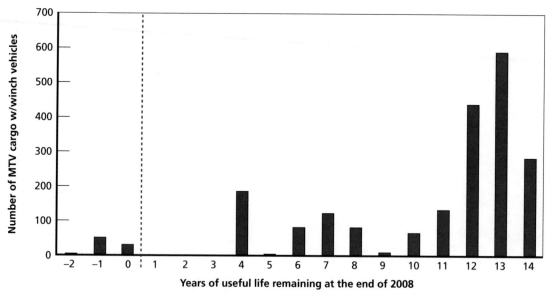

NOTE: Assumes 15-year EUL.

RAND *TR890-B.8*

MTV Air Drop Cargo Fleet Profile

Figure B.9
MTV Air Drop Cargo Fleet Profile

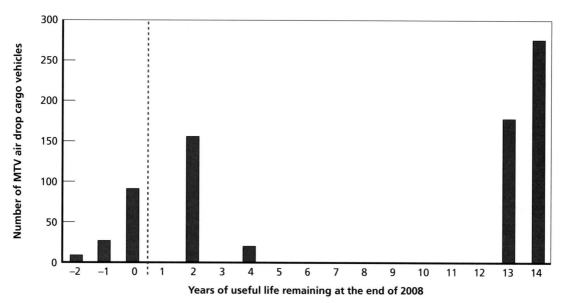

NOTE: Assumes 15-year EUL.
RAND *TR890-B.9*

MTV Air Drop Cargo with Winch Fleet Profile

Figure B.10
MTV Air Drop Cargo with Winch Fleet Profile

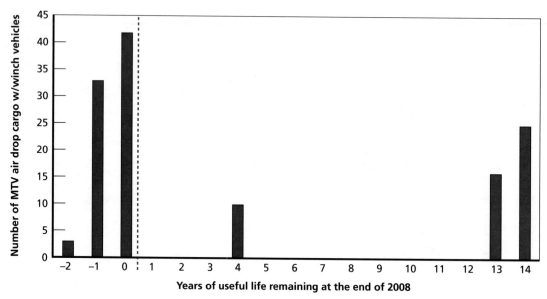

NOTE: Assumes 15-year EUL.
RAND *TR890-B.10*

MTV Long Wheel Bed Cargo Fleet Profile

Figure B.11
MTV Long Wheel Bed Cargo Fleet Profile

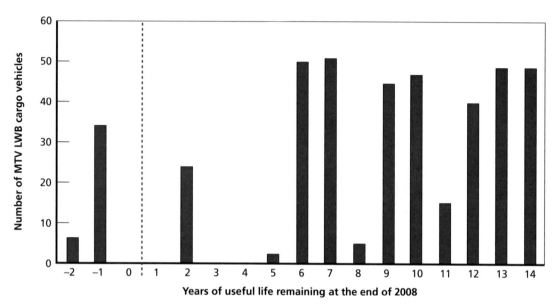

NOTE: Assumes 15-year EUL.
RAND *TR890-B.11*

MTV Long Wheel Bed Cargo with Winch Fleet Profile

Figure B.12
MTV Long Wheel Bed Cargo with Winch Fleet Profile

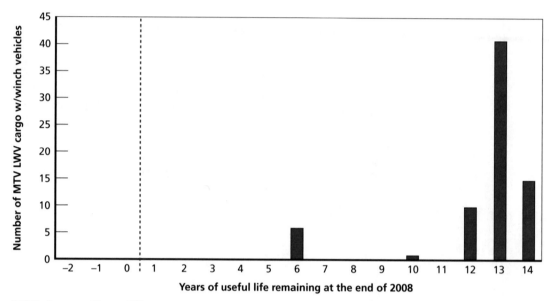

NOTE: Assumes 15-year EUL.
RAND *TR890-B.12*

MTV Cargo with Man Handling Equipment Fleet Profile

Figure B.13
MTV Cargo with Man Handling Equipment Fleet Profile

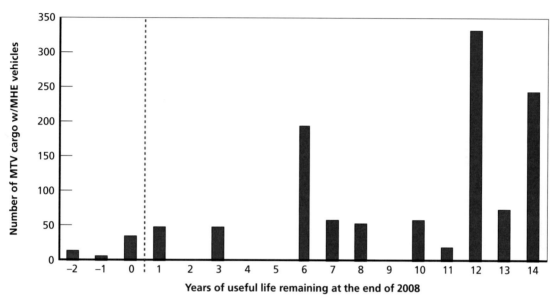

NOTE: Assumes 15-year EUL.

RAND *TR890-B.13*

MTV Long Wheel Bed Cargo with Man Handling Equipment Fleet Profile

Figure B.14
MTV Long Wheel Bed Cargo with Man Handling Equipment Fleet Profile

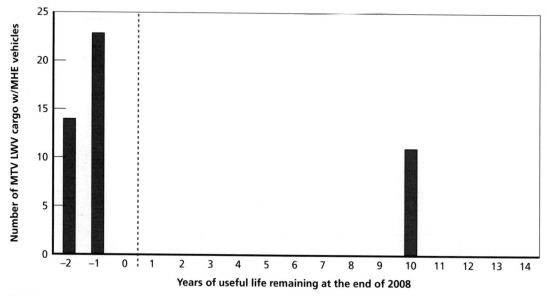

NOTE: Assumes 15-year EUL.
RAND *TR890-B.14*

MTV Dump Truck Fleet Profile

Figure B.15
MTV Dump Truck Fleet Profile

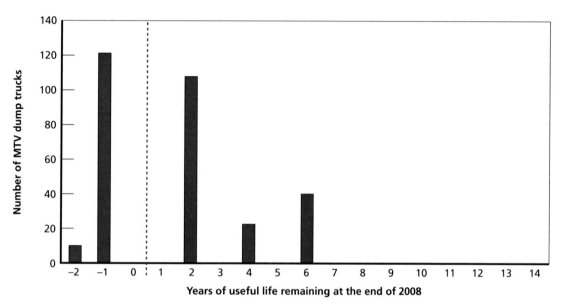

NOTE: Assumes 15-year EUL.
RAND *TR890-B.15*

MTV Dump Truck with Winch Fleet Profile

Figure B.16
MTV Dump Truck with Winch Fleet Profile

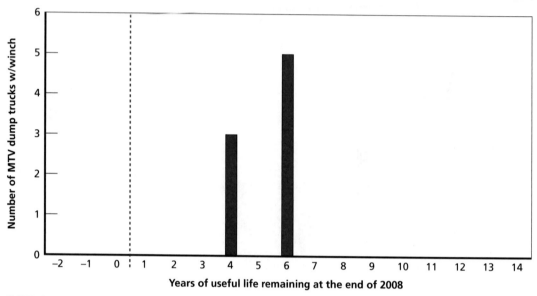

NOTE: Assumes 15-year EUL.
RAND *TR890-B.16*

MTV 10-Ton Dump Truck with Winch Fleet Profile

Figure B.17
MTV 10-Ton Dump Truck with Winch Fleet Profile

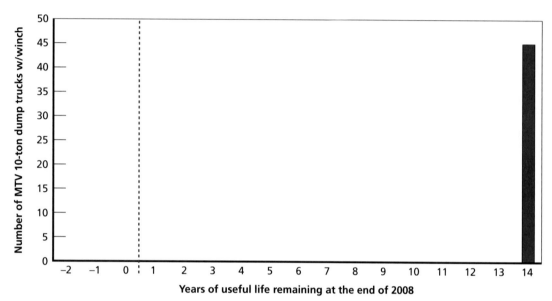

NOTE: Assumes 15-year EUL.
RAND *TR890-B.17*

MTV Tractor Fleet Profile

Figure B.18
MTV Tractor Fleet Profile

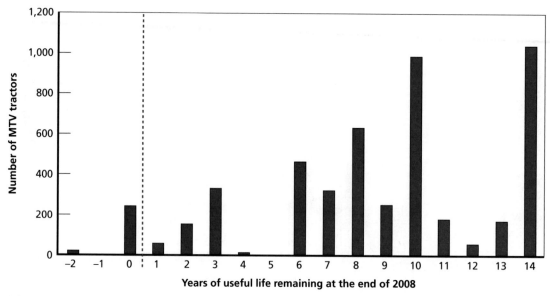

NOTE: Assumes 15-year EUL.
RAND *TR890-B.18*

MTV Tractor with Winch Fleet Profile

Figure B.19
MTV Tractor with Winch Fleet Profile

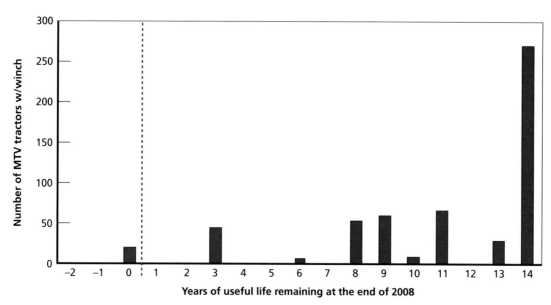

NOTE: Assumes 15-year EUL.
RAND *TR890-B.19*

MTV Wrecker with Winch Fleet Profile

Figure B.20
MTV Wrecker with Winch Fleet Profile

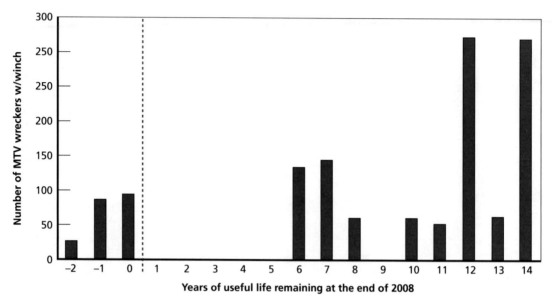

NOTE: Assumes 15-year EUL.
RAND *TR890-B.20*

MTV EXP Van Fleet Profile

Figure B.21
MTV EXP Van Fleet Profile

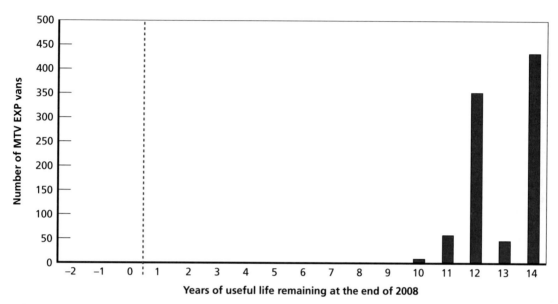

NOTE: Assumes 15-year EUL.
RAND *TR890-B.21*

MTV Load Handling System Truck Fleet Profile

Figure B.22
MTV Load Handling System Truck Fleet Profile

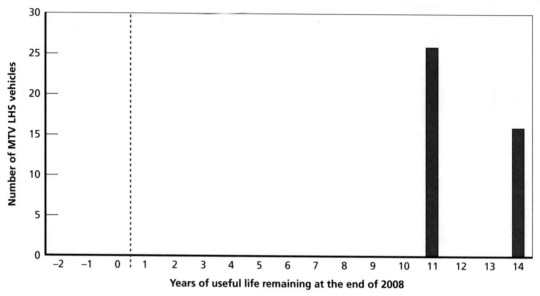

NOTE: Assumes 15-year EUL.
RAND *TR890-B.22*

MTV Chassis Fleet Profile

Figure B.23
MTV Chassis Fleet Profile

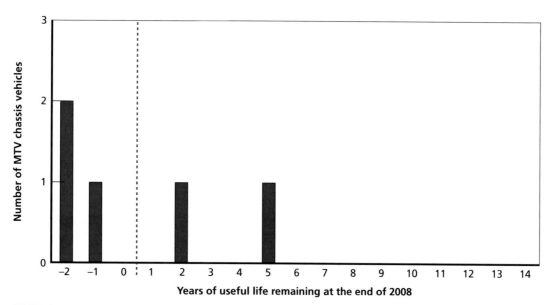

NOTE: Assumes 15-year EUL.
RAND *TR890-B.23*

MTV Long Wheel Bed Chassis Fleet Profile

Figure B.24
MTV Long Wheel Bed Chassis Fleet Profile

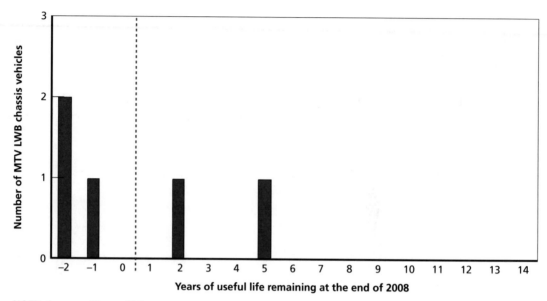

NOTE: Assumes 15-year EUL.
RAND *TR890-B.24*

MTV High Mobility Artillery Rocket System Resupply Vehicle Fleet Profile

Figure B.25
MTV High Mobility Artillery Rocket System Resupply Vehicle Fleet Profile

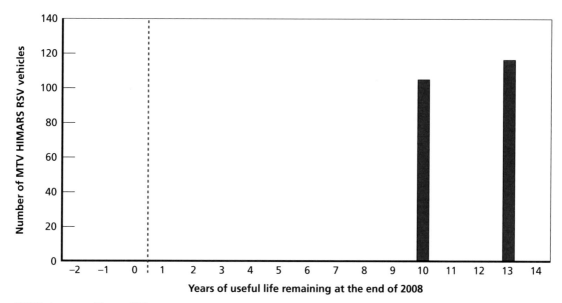

NOTE: Assumes 15-year EUL.

RAND *TR890-B.25*

Examination of the Light TWV Fleet

In FY 2007, RAND Arroyo Center performed a short six-week effort focused on light tactical vehicles. This short-term effort examined the Army's light tactical vehicle fleet to identify conditions that would induce the Army to make resource decisions related to the High Mobility Multipurpose Wheeled Vehicle (HMMWV), the Mine Resistant Ambush Protected (MRAP) vehicle, and the Joint Light Tactical Vehicle (JLTV). Production data on the light tactical vehicle (LTV) fleet was provided to the study team by the LTV Integrated Product Team, and the light fleet analysis was based on that production data.

Examination of the LTV fleet revealed that in FY 2007, approximately 40 percent was beyond its expected useful life.[1] Moreover, operations and maintenance, Army (OMA) costs would continue to increase as this fleet aged. Approximately 55,000 model A0 and A1 light vehicles that are beyond useful life can be recapitalized. In the future, the useful life and size of the LTV fleet can be maintained by purchasing, on average, approximately 7,000 vehicles per year.

In FY 2007, 78 percent of the LTV fleet was unarmored. Purchasing MRAP vehicles at a cost of approximately $1 million each will achieve the immediate force protection goal and might mitigate the Long Term Armoring Strategy requirement. In the longer term, MRAP provides a strategic hedge as a theater reserve for future contingencies. These MRAP vehicles can also be integrated into the fleet with special roles such as explosive ordnance detection (EOD), ambulance service, and command and control (C2).

The 2007 effort also found that, until the decision is made that no new Up Armored HMMWV (UAH) will be procured, the UAH production line must remain open. The minimal efficient production rate for that line is 3,756 vehicles per year.[2] Buying a mixture of JLTV and UAH would maintain both fleet size and expected usefulness. There is also an opportunity for an HMMWV Product Improvement Program (PIP).

Profiles of EUL for the UAHs and model A0 and model A1 HMMWVs show that there are waves consisting of 36,000 UAHs and 92,000 model A0/A1 HMMWVs that will reach the end of their EUL between 2016 and 2024.

---

[1] In the examination of the light vehicle fleet, the years of expected useful life for a vehicle class is defined as the planned number of years a vehicle is expected to be mission capable without incurring expenses exceeding the cost of replacement or recapitalization. While vehicles should not be discarded simply because they have reached the end of EUL, the concept is a good planning tool to develop recapitalization and replacement strategies. The EUL concept can also indicate when a vehicle might no longer have the technological features to perform Army military missions. (See Chapter Two of this report for further elaboration of the EUL concept and its relationship to other equipment wear-out concepts.)

[2] Production rate and other data used for this analysis were furnished by G-8, LTV Force Development IPT. Current as of June 19, 2007, these data were accepted on face value and not verified as part of this study.

LTV Capabilities Comparison

We compared the HMMWV, UAH, JLTV, and MRAP with respect to three capabilities: (1) mine and underbody improvised explosive device (IED) protection, (2) performance (mobility), and (3) payload (measured in pounds). Our evaluations are based on a U.S. Marine Corps and Army comparison study of the HMMWV, MRAP, and JLTV.[3]

The HMMWV has virtually no mine and IED protection, but has good mobility and payload characteristics and is the least expensive of the four vehicles. The UAH with Fragmentation Kit 5[4] has better mine and underbody IED protection than the HMMWV, but that protection is still poor and the added armor reduces the vehicle's mobility and payload. The UAH is more expensive than the HMMWV due to the added cost of the armor. By contrast, the JLTV has good mine and IED protection, good mobility, and good payload, but costs more than the UAH. Finally, the MRAP has excellent protection and outstanding payload, but has very poor mobility and is very expensive at about $1 million per vehicle. However, anecdotal evidence indicates that the added protection has resulted in tangible benefits: BG John Allen, deputy commander of coalition forces in Anbar province, stated that IED attacks on MRAPs result in less than half the casualties of IED attacks on UAHs.[5] Figure C.1 illustrates these comparisons.

Findings from LTV Examination

The primary finding of the LTV examination is that as of June 19, 2007, the HMMWV and MRAP strategy that the Army has in place will not meet the Grow the Army (GTA) (145,000 vehicles) and Long Term Armoring Strategy (LTAS) (42 percent armored) goals unless the Army alters its strategy. The study team assumed a 1 percent noncombat attrition rate for the light fleet analysis. Estimates from the LTV examination show that in 2015, the Army will have 131,000 vehicles of A0/A1s, recapitalized A0/A1s, A2s, or UAH in garrison. Of these 131,000 vehicles, only 38 percent will be armored, well short of the Army's LTAS of 42 percent armored. Not counting the additional 21,300 M1114s and M1151s in theater, the 131,000 vehicles also fall short of the 145,000-vehicle GTA requirement. These findings are illustrated in Figure C.2.

In our analysis, we assumed that a recapitalized vehicle will gain ten additional years of EUL from the date of recapitalization. A new UAH will have a EUL of fifteen years.[6] With the planned limited UAH buy and the recapitalization strategy in place as of June 19, 2007, the Army cannot maintain the GTA[7] requirement of 145,000 vehicle LTVs with an average age below the EUL. Assuming a 1 percent attrition rate for vehicles, an acquisition rate of

---

[3]   CPT Jeremy Gray, *Tactical Wheeled Vehicle Armor Classification System*, SECRET/NF//20161030, sanitized by Walter Nelson, RAND Classified Library Services Manager, November 6, 2008.

[4]   An example of a Fragmentation Kit 5 can be found in BG John R. Bartley, Program Executive Officer, CS & CSS, "2005 Advanced Planning Brief to Industry," October 28, 2005, p. 23.

[5]   Tom Vanden Brook, *USA Today*, "New vehicles protect Marines in 300 attacks in Iraq province," April 19, 2007, p. A4.

[6]   The ten additional years of useful life from year of recapitalization figure and the fifteen years of useful life figure for new vehicles were furnished by the LTV IPT and were current as of June 19, 2007.

[7]   Army Growth Plan, December 19, 2007.

Figure C.1
Comparison of LTV Capabilities

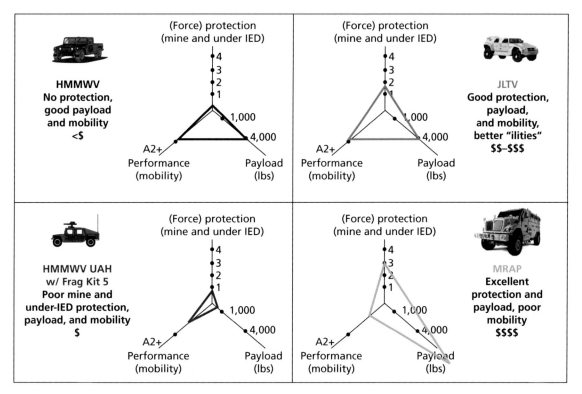

NOTE: Comparisons based on following source: CPT Jeremy Gray, *Tactical Wheeled Vehicle Armor Classification System*, SECRET/NF//20161030, sanitized by Walter Nelson, Classified Library Services Manager, November 6, 2008.
RAND *TR890-C.1*

6,600 new and recapitalized vehicles per year is required to maintain such a fleet. In addition, our analysis showed that the estimated costs of maintenance and downtime over ten years assuming only 6 percent of the fleet is non-mission-capable on average is $0.75 billion. With a more realistic assumption that, on average, 10 percent of the vehicles in the fleet are non-mission-capable, the maintenance and downtime costs over ten years rise to $3.24 billion, and a 16 percent assumption will entail maintenance and downtime costs of $7.27 billion. These maintenance and downtime cost estimates suggest that the Army should consider alternative acquisition and recapitalization strategies. Since there are already 55,000 A0/A1s in the fleet that have exceeded their EUL as of June 19, 2007, the potential benefits of pursuing such strategies will be greater the sooner the Army can take action on revising strategies. Figure C.3 illustrates these findings.

Figure C.2
2007 HMMWV and MRAP Strategy Will Not Meet GTA and LTAS Requirements by 2015

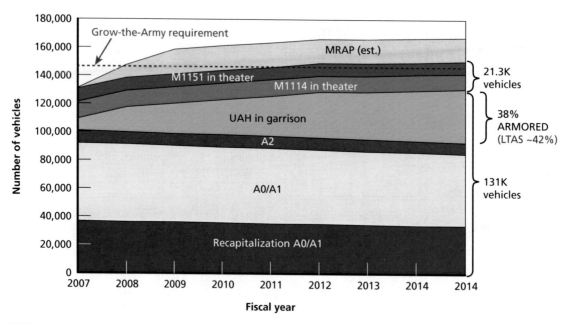

NOTE: 1% per year noncombat attrition rate included.
RAND *TR890-C.2*

Figure C.3
EUL Profile of LTV Fleet as of June 2007

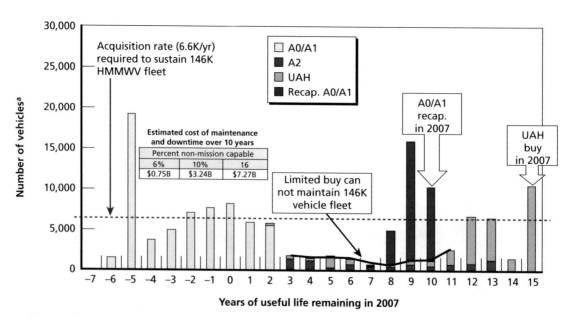

[a]Includes vehicles in theater.
NOTES: Assumes currently planned UAH buy and recapitalization strategy. Estimated additional costs of
A0/A1 fleet over 10 years without recapitalization.
RAND *TR890-C.3*

Improving the Army's Knowledge Base

The ability to develop, update, and maintain a data base sufficient to provide the Army with profiles that accurately reflect the actual status of its TWV fleets is dependent on having accessible and reliable data. This appendix outlines some challenges encountered during the course of the study and offers candidate avenues that might mitigate such trials for future data investigations.

Identification of Data Sources

It is reasonable to expect that research will entail some effort to identify appropriate data sources. The experience in this study was not an exception. Our investigations did not reveal an Army data base that allowed for readily deriving the status of the Army's medium and heavy TWV fleets from a single source.

One avenue for facilitating the task of updating the status profiles generated in the study is for the Army to maintain a meta-data base that lists the name of the data base, the agency that maintains the data base, and a short description of the purpose, use, contents, currency, accessibility, and point of contact for further inquiries. Such a source would greatly simplify data identification tasks for all potential future users of the Army's data repositories. The current informal process that relies on corporate knowledge will always be a part of the data source identification task, but a meta-data base can ensure more thorough, more effective, and more efficient execution and, thus, superior results in a shorter amount of time.

Access to the Data Bases

Once access to various Army data bases was attained, the study team found it was difficult to ascertain which data bases should be used for what purposes. Guidance toward extracting relevant material was largely absent, and permissions and query inputs were not transparent. Denial of access and unknown biasing of output were consequences of the lack of transparency. The meta-data base suggested above could help alleviate some of these concerns. Readily accessible users' guides would alleviate the majority of access concerns.

The Data in the Data Bases

The contents of the data bases were often voluminous, but it was unclear how accurately the data reflected reality. For example, downloads labeled "on hand" are interpreted as vehicles in the data base. The relationship between the "on hand" quantities and vehicles physically present at an Army site or in use by Army personnel is not known.

Data were sometimes incomplete in the sense that the data were missing with no clear guidance on interpretation. For example, blanks could be interpreted as unknown, unrecorded, zero, not available, or inapplicable. For the purposes of this study, large banks of blanks required careful interpretation. For instance, odometer readings could be blank because the vehicles have been replaced, so the proper interpretation would be "inapplicable" and the vehicles should not be included in quantities of physically existing vehicles. In other cases, the records may simply have not been updated yet, so the proper interpretation of such blanks would be "unrecorded," but the vehicles still physically exist and are part of the inventory count.

The currency of the data was not always apparent. For example, date of last update was often not available. In addition, the original sources of the data or party responsible for recording the data were sometimes not stated.

Finally, the formulas and rules used to derive outputs from inputs were not described. This absence limited the user's ability to interpret the outputs.

The degree of internal data consistency among the data bases was unclear. As might be expected, the formats varied widely among data bases, so direct comparisons were difficult. For example, budget data shows procurements and manufacture schedule by contract, which could span any number of years or fiscal years. Form DD250s show acceptance per vehicle by date. PM and contractor data were often available only in aggregate by model or other schemes that might not indicate any date or span of time.

No data base is "perfect" for any use, so it is unrealistic to expect no challenges, but more thorough and available documentation about the data in the data bases would provide more transparency as well as allow for more accurate interpretation of data extractions. Such improvements could lead to more accurate findings. The keepers of the data bases would need to balance the benefits of such additional documentation with the cost and effort required to make it available.

Bibliography

Army Inflation Indices per Army FY09 Infl_30Jan2008.xls.

Communications with LTV IPT, March–June 2007.

Communications with TWV IPT, October 2007–September 2008.

Estimated Trailer Production by Year Slides, furnished by TWV IPT, April 2008.

Gray, CPT Jeremy, *Tactical Wheeled Vehicle Armor Classification System*, SECRET/NF//20161030, sanitized by Walter Nelson, Classified Library Services Manager, November 6, 2008.

HEMTT Recapitalization Analysis Workbook, furnished by TWV IPT, April 2008.

Johnson, A., and M. Scharra, *Weapon Systems Review POM 10-15*, November 15, 2007.

Line Haul Funding 2005–2008 Workbook, furnished by TWV IPT, April 2008.

Line Haul Trailer Fleet Statistics Workbook, furnished by TWV IPT, April 2008.

Selected Acquisition Report (SAR), December 2007.

Tactical Wheeled Vehicle (TWV) Protection Level TPE Workbook, December 3, 2007.

Tank-Automotive and Armaments Command (TACOM) Fleet Assessment Team, *The Tactical Vehicle Fleet Book*, Tactical Vehicle Product Support Integration Directorate, February 1, 2006.

U.S. Army Procurement Programs, "FY 1996/1997 Biennial Budget Estimate: Other Procurement, Army, Activity 1, Tactical and Support Vehicles," *Committee Staff Procurement Backup Book*, February 1995.

———, "FY 1997 Budget Estimate: Other Procurement, Army, Activity 1, Tactical and Support Vehicles," *Committee Staff Procurement Backup Book*, March 1996.

———, "FY 1998/1999 Budget Estimate: Other Procurement, Army, Activity 1, Tactical and Support Vehicles," *Committee Staff Procurement Backup Book*, February 1997.

———, "FY 1999 Budget Estimate: Other Procurement, Army, Activity 1, Tactical and Support Vehicles," *Committee Staff Procurement Backup Book*, February 1998.

———, "FY 2000/2001 Budget Estimate: Other Procurement, Army, Activity 1, Tactical and Support Vehicles," *Committee Staff Procurement Backup Book*, February 1999.

———, "FY 2001 Budget Estimate: Other Procurement, Army, Activity 1, Tactical and Support Vehicles," *Committee Staff Procurement Backup Book*, February 2000.

———, "FY 2002 Amended Budget Submission: Other Procurement, Army, Activity 1, Tactical and Support Vehicles," *Committee Staff Procurement Backup Book*, June 2001.

———, "FY 2003 Budget Estimate: Other Procurement, Army, Activity 1, Tactical and Support Vehicles," *Committee Staff Procurement Backup Book*, February 2002.

———, "FY 2004/2005 Biennial Budget Estimate Submission: Other Procurement, Army, Activity 1, Tactical and Support Vehicles," *Committee Staff Procurement Backup Book and Multiyear Exhibits*, February 2003.

————, "FY 2005 Budget Estimates: Other Procurement, Army, Activity 1, Tactical and Support Vehicles," *Committee Staff Procurement Backup Book*, February 2004.

————, "FY 2006/2007 Budget Estimates: Other Procurement, Army, Activity 1, Tactical and Support Vehicles," *Committee Staff Procurement Backup Book*, February 2005, pp. 48–53.

————, "FY 2007 Budget Estimates: Other Procurement, Army, Activity 1, Tactical and Support Vehicles," *Committee Staff Procurement Backup Book*, February 2006.

————, "FY 2007 Supplemental Budget Estimate: Other Procurement, Army, Activity 1, Tactical and Support Vehicles," *Committee Staff Procurement Backup Book*, February 2007a.

————, "FY 2008 Grow the Army Detail: Other Procurement, Army, Activity 1, Tactical and Support Vehicles," *Committee Staff Procurement Backup Book*, February 2007b.

————, "FY 2009 Budget Estimates: Other Procurement, Army, Activity 1, Tactical and Support Vehicles," *Committee Staff Procurement Backup Book*, February 2008.

Vanden Brook, Tom, *USA Today*, "New vehicles protect Marines in 300 attacks in Iraq province," April 19, 2007, p. A4.